Basic Principles of Automatic Control

Basic Principles of Automatic Control

With special reference to Heating and Air Conditioning Systems

W. H. Wolsey FIHVE

Hutchinson Educational

Hutchinson Educational Ltd
3 Fitzroy Square, London W1

London Melbourne Sydney Auckland
Wellington Johannesburg Cape Town
and agencies throughout the world

First published 1975

This book has been set in IBM Press Roman by E.W.C. Wilkins Ltd., London and Northampton and printed in Great Britain on smooth wove paper by Anchor Press Ltd and bound by Wm Brendon & Son Ltd both of Tiptree, Essex

ISBN 0 09 115490 1

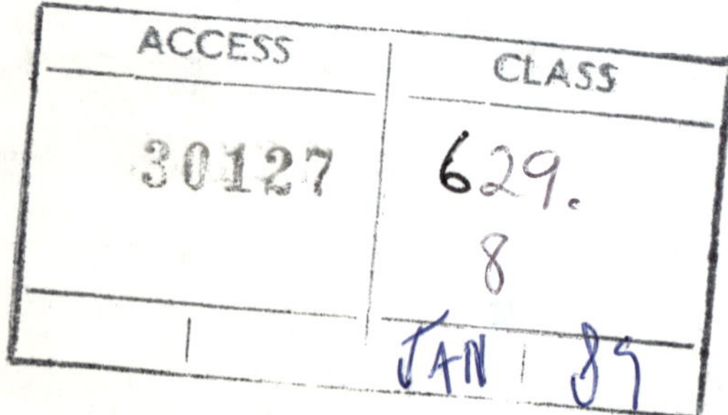

Contents

Symbols, Units and Abbreviations

Symbol	*Unit*	*Item, Quantity or Meaning*
A	–	Action also double the amplitude of an oscillating variable expressed in terms of that variable.
A	m^2	Area
a	%	Authority of an auxiliary detector
a_L	%	Authority of a limiting device
a_p	various	Process amplification $d\gamma_0/ds$
B	–	Boost, branch
C	–	Calorifier, relay coil
C	F,m^2, etc.	Capacity
c	–	Constant
Cal	–	Adjustment for calibration purposes
D	–	Sensing element
D	–	When applied to a contact, a position denoting a demand (e.g. for more heat)
d	m, mm	Distance
E	–	Exhaust
F	–	Flapper
f	s^{-1}	Frequency
FB	–	Feedback
FCE	–	Final control element
G	–	Gain
H	J/s	Heat flow
h	m	Height
HA	–	Heat acceleration
HL	–	High limit
I	A, mA	Electric current
i	%	Impulse as a fraction of t_0
K_1	–	Proportional action factor
K_2	–	Integral action factor
K_3	–	Derivative action factor

K_4	$m^3/bar^{\frac{1}{2}}$, h	Valve capacity factor
L	V	Line voltage
L	m	Length, thickness
LL	–	Low limit
M	p.s.i., bar	Mains pressure in pneumatic control schemes
N	–	Neutral, nozzle
n	s^{-1}	Number of revolutions per second
NSB	°C	Night set-back
p	bar, p.s.i.	Pressure
p.c.c.	–	Process control characteristic
pH	–	Acidity
Q	kJ/h,%	Flow of heat
q	m^3/s	Quantity of water, steam etc. flowing per unit time
q	Coulomb	Electric charge
R	s	Relay restriction
R	Ohm, $m^2h°C/kJ$	Resistance, electrical or thermal
R_{s_1}, R_{s_2}	m^2,h,°C/kJ	Internal, resp external thermal surface resistance
R_d	various	Discharge resistance
r	$m^2h°C/kJm$	Specific resistivity e.g. of a wall
rH	%	Relative humidity
S	–	Spring i.e. an electric device; summer; when applied to a contact, a position denoting that the demand has been satisfied; signal
S	various	Step change in various types of floating controller
Sp	various	Span i.e. control range, i.e. the difference in γ_0 at $s = 0$ and 100% respectively
s	%	Movement of final control element as a percentage of its stroke
T	°K	Absolute temperature
T, T_D	s, sec	Derivative action time
T, T_I	s, sec	Integral action time
T_1	–	Main temperature sensing element
T_2	–	Snubber thermostat
T_3	–	Outside, compensating thermostat
T_{cr}	s	Oscillation time of a P controller adjusted critically

T_d	s	Distance – velocity lag, dead time
T_{de}	s	Effective dead time
Th	s	Half life period
T_t	s	Transfer lag
t	s	Time
t_0	s	Impulse time
U	$kJ/m^2h°C$	Overall thermal transmittance coefficient
u	various	Disturbance
V	–	Valve
V	various	(Controller-or Bridge-) output signal
V	m^3	Volume
V	V	Voltage
W	–	Winter
α,β	angular degrees	Angle
γ,γ_t	various	Deviation, at time *t*
γ_c	various	Range of values to which a controller can be adjusted
γ_d	various	Differential of a two-step controller
γ_i	various	Desired value
$\gamma_0,\gamma_1,\gamma_2$	various	Controlled condition at time 0, 1, 2, etc.
γ_{max}	various	Amplitude
γ_n	various	Neutral zone
γ_p	various	Proportional band or throttling range
Δ	–	Fraction, small change
θ	°C	Temperature
θ_a	°C	Ambient temperature
θ_e	°C	Effective temperature
θ_0	°C	Outside temperature
θ_r	°C	Radiation temperature
τ	s	Time constant, associated with the process controlled, exponential lag
τ_c	s	Exponential lag introduced by the controller
φ	%	Relative humidity
ψ	various	Load

Preface

Many an engineer confronted with automatic control finds that the problems concerned go far beyond those for which his training had prepared him. For one whose student days lie in the past, the study of automatic control theory is daunting. Not only will he have to refresh his knowledge of mathematics, but he will also have to tackle several mathematical subjects he is unlikely to have encountered before.

In reading this book he will find manyof the more important concepts of automatic control treated in a simple manner.

Thus the reader before entering on his main study, will have developed a feeling for the subject and knows what lies behind the rather abstruse science of automatic control. Should he later become discouraged, the many practical examples given in this book will at least have provided him with sufficient background to be able to deal with many of the more simple control problems he is likely to meet.

Automatic control is applied in numerous specialised fields of technology, the practice of which, though simple enough to the group of engineers concerned, is a complete mystery to most others. Thus the selection of examples with which the text is illustrated was not an easy matter.

In choosing these from the field of heating, ventilating and air-conditioning, it is hoped that a subject has been found with which many engineers will be more or less familiar. This field is peculiar in as much as the use of automatic controls is so wide spread that several firms have developed specialised equipment with such applications in mind. Such equipment is produced in quantities far surpassing that for any other field. Among the sales staff there are many engineers with a detailed knowledge of the subject, well qualified to give advice in most cases. Thus the heating and air-conditioning engineer, when in need, can call upon expert advice. Yet, as many will wish to retain their independence, the text has been amplified to deal with subjects with which he alone is concerned, the author has tried to compile a book that while intelligible to engineers in general gives the heating and air-conditioning engineer most of the theoretical knowledge he is likely to require in his day-to-day practice.

Acknowledgements

The subject matter presented here, evolved over a long period while I was in charge of training young engineers wishing to join one of several European subsidiaries of Satchwell Control Systems Ltd., a firm with which I have been associated for many years.

I am especially indebted to the parent company in Slough, a member of the GEC Electric Components Group, Thermal Controls Division, for allowing part of the preparation of this book to be carried out in the company's time; and for providing suitable cover illustrations.

In the preparation of this work I have benefited greatly from the generous technical advice of D.J. Croome-Gale, of the University of Technology at Loughborough.

Finally, my thanks to my publishers for their care in the planning and execution of this book.

1. The process as part of the control loop

1.1. The control engineer's vocabulary

The arrangement by which a machine or a process is made to control itself without human interference, other than the initial setting of the necessary devices and possibly an occasional check, is not new; James Watt constructed his governor nearly two centuries ago. Nevertheless, labour being cheap and processes relatively simple, progress in automatic control was slow; not until shortly before and during the Second World War were its full potentialities realised. The theory was systematically developed and is now a discipline covering a field wide enough to justify special departments at most technical colleges and universities.

As a result, a whole new vocabulary has come into being which allows control engineers to describe the concepts with which they deal without ambiguity. Some concepts, taken from other fields, have somewhat changed their meaning, and associations of this kind may cause confusion. It would therefore seem logical to devote a short introductory section to an explanation of some of the more common control terms and also such symbols as have been standardised. Most of these are listed in *British Standard 1523*, a 'Glossary of Terms used in Automatic Controling and Regulating Systems', parts 1 and 5 having special reference to process and kinetic control, revised in 1967.[1,2] An American list of standardised terms is also available.[3]

As a convenient example the simple central heating installation illustrated in Fig.1.1 may serve. The *controlled condition* γ_0, i.e. the physical condition of the process it is wished to keep constant or within close limits, is in this case the space temperature θ_{sp}, in the room. The value at which the space temperature is maintained, i.e. the *desired value* or reference input γ_i, is in this case the 20°C to which the thermostat in the room has been adjusted. Inasmuch as the boiler temperature θ_b is also kept within certain limits, the figure shows a second process independent of the first. The fact that the two controlled conditions here are both temperatures, is incidental. In other cases the same symbol

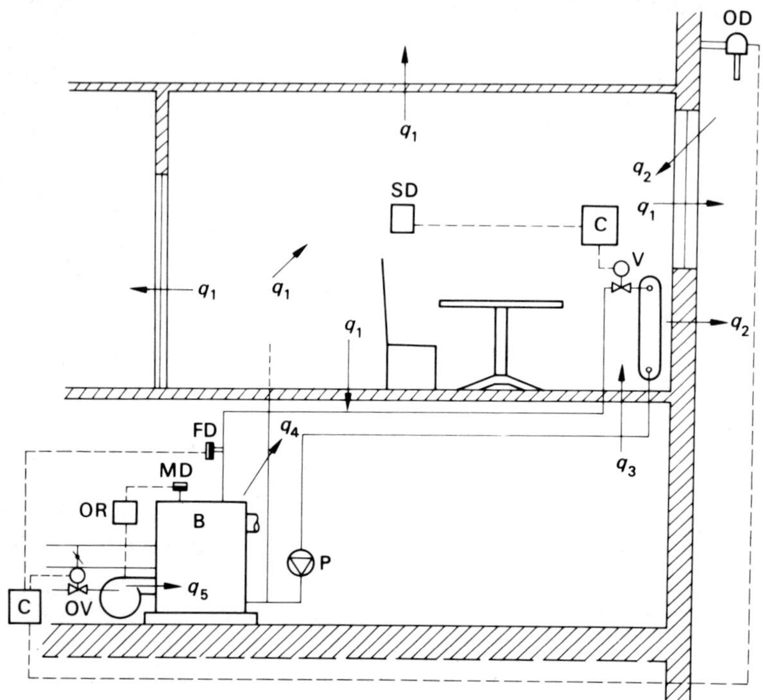

Fig. 1.1
A simple control scheme in which the ambient temperature in a room is kept at a constant value or at least within acceptable limits. FD, MD, SD, OD, flow, maximum, space and outside temperature sensing element respectively; *V*, control valve; q_1, heat loss through walls; q_2, q_3, heat gain through solar radiation and through floor; q_4 heat loss from B, boiler; q_5 heat from oil burner; C, controller, OR, oil burner relay; P, pump, OV oil valve

γ_0 could be used to indicate a pressure p; a pressure difference Δp; a liquid level h; a relative humidity φ; a degree of acidity, pH; a number of revolutions per second n; or any other feature or condition of the process that should be kept at a constant value (or as nearly constant as possible).

The two separate *processes* in our example, the heating of the room – including its walls and the air in the room – and that of the boiler and its contents that are continually changing, inasmuch as the flow extracted from the boiler is made up by the return. *B.S. 1523* defines process term as: ‘the act of physically or chemically changing matter or converting energy'. A distinction is made between process and *plant*, the latter being the installation in which the process is carried out. The plant installed in order to keep the boiler flow temperature constant consists of the oil burner, its relay etc. and the boiler, whereas the plant for the space heating process includes these items as well as the pump, flow and return mains the control valve and the space heating appliances.

The regulating system by which the control is carried out has the following main features: firstly the magnitude of the controlled condition is measured by a *sensing element* and compared with the value it is wished to maintain – the desired value γ_i. If the two are not identical, the comparing element emits a signal θ, reflecting both character and, usually, the magnitude of the deviation γ. At this stage, the signal is generally expressed in units differing from those of the controlled variable. In the example, the deviation, measured in degrees, has been converted to an electrical voltage, partly because measuring and comparing can conveniently be carried out that way, and partly because after having been processed if necessary and amplified, the same signal can be used to energise the motor driving the valve or whatever final control element happens to be used.

The final control element now changes the flow of matter or energy to the process in such a way that the deviation is reduced, if possible, to zero. The change in γ is measured once more, the signal modified, the position of the final control element changed, and so on. The system forms a closed loop in which each element, receiving a signal from the preceding one processes it and passes it on to the element following it. Figure 1.2 shows the standard designation of the signal in consecutive stages of its development as it moves round the control loop. If such detail is not necessary, the output signal of each individual element may also be indicated by the symbol V.

The most important feature of such a closed-loop system is the fact that it is continuously monitored by the process itself. The sensing element, (i.e. the element which initiated the signal that set the control system in motion) is in a position to check the result of its action and to correct it where necessary, until it has reached the smallest deviation value the control system is capable of achieving. The *final control element* may be a whole plant in itself, such as a turbine driving a rotary compressor in a cooling plant, a boiler complete with burner and ancillary equipment. It may also be a simple pump or, more usually, a valve or damper positioned by an electric or pneumatic *actuator*. Though schemes exist by which an electric signal is, in the final stage, converted

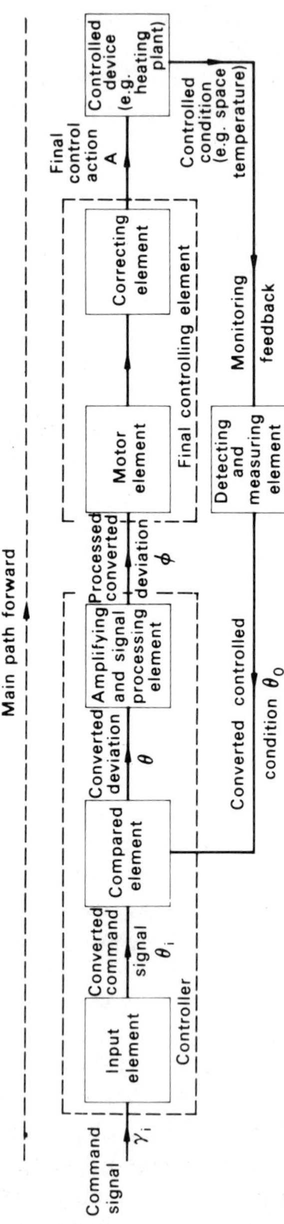

Fig. 1.2
Block diagram of the space temperature control system of Fig. 1.1. The standard symbols for the signal as it passes from one stage of the control circuit to another are also indicated.

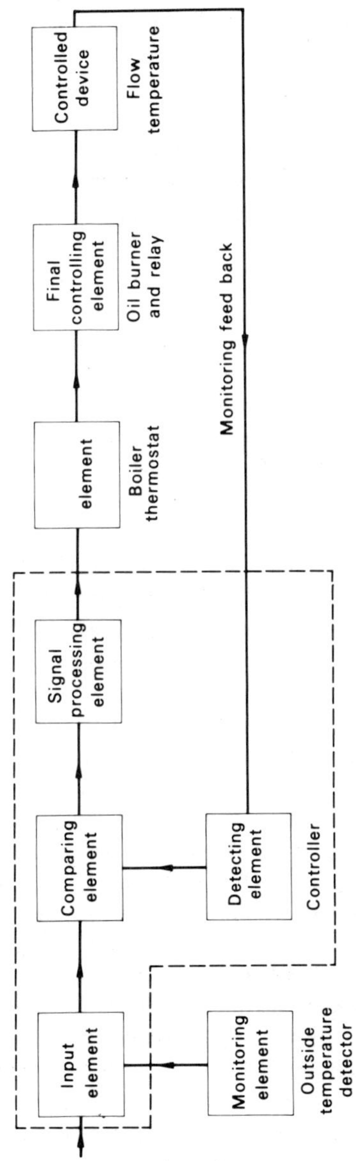

Fig. 1.3

As Fig. 1.2 for the flow temperature control system containing an open-loop monitoring element in the form of an outside temperature detector.

to an air pressure suitable for activating a pneumatic motor, it is usually the motor element that determines the kind of signal to which the deviation is converted by the controller. Thus there are electric, electronic, pneumatic and hydraulic control systems in which all elements are usually designed to produce signals uniform in nature and distinctive to the particular type of controller.

As stated, the temperature control systems for space and boiler of our example both form closed loops. Each thermostat (i.e. temperature sensing element) not only directs the operation of its respective final control element, i.e. the valve and the oil burner, but it is also exposed to the effect of the action it has initiated and will modify its signal as required.

In the case of the boiler there is a further complication. Although the thermostat setting corresponds to a certain value, this is not necessarily the desired one. The latter is raised as the outside temperature drops and vice-versa; the desired value therefore rises and falls as the weather changes. To this end a thermostat *OD*, the monitoring element of the system, is mounted on an outside wall. It measures the outside temperature, and the controller, receiving its signals, reduces the desired value which *FD* tries to maintain as the ambient temperature rises. The object of this scheme is to adjust the flow temperature to a value such that, if the flow reaches a radiator without further controls being interposed, the heat emitted will more or less conform to requirements.

Whether the scheme fulfills its purpose cannot be detected by *OD*; this could only be ascertained by measuring the resulting space temperature which *OD* is not in a position to do. Moreover, in the absence of a monitoring feedback loop, local disturbances are not taken into account; this therefore, is an example of an *open-loop element*. The scheme is illustrated in the block diagram of Fig. 1.3.

A more detailed description of the instruments used may help to clarify what happens. The space temperature is measured by the sensing element SD, which converts this temperature to a change in voltage. The controller receives the signal and compares it to the value to which it was set. It then amplifies the difference between the two and, in its turn, passes a signal to the final control element, the valve. This changes its position so as to allow more or less heat to be transferred to the space according to whether the deviation was negative or positive. Thus the deviation will be reduced, and the valve position which, in order to get quick results, had initially been changed rather more than was strictly necessary, will be modified once more in response to the effect of its own action.

This interaction between controlled variable and controller, described as a series of steps, is actually a *continuous* process. Though the standard does not mention the term, it is generally called a *modulating control system* thus distinguishing it from the *discontinuous* two-step, (or on-off) or multi-step regulating system controlling the burner.

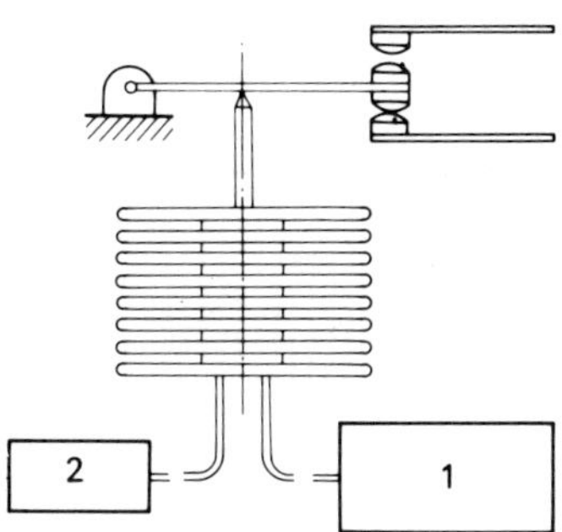

Fig. 1.4
An obsolete type of compensator showing how the open and closed-loop detectors co-operate to achieve the shift in desired value.

To illustrate this, an obsolete type of 'weather controller' (Fig. 1.4) has been chosen as the various features of the system are more easily understood with this device. *OD* and *FD* are each equipped with a phial containing a liquid which expands and contracts with temperature fluctuations. Both phials are connected through capillaries to bellows which control a contact placed in the control circuit of the oil burner. As the flow temperature drops below the temperature to which the device is set, the liquid in the *FD* phial contracts, withdraws fluid from the bellows which, on contracting, closes the lower contact and switches the oil burner on. As a result, the flow temperature rises, and some liquid is expelled from the *FD* phial to the bellows which start to expand. At a given temperature the contact is broken and the flame in the boiler is extinguished. This is the closed-loop two-step control system in which the valve as final control element has been replaced by the boiler plant. The action of this two-step control system is discontinuous: the oil burner is either in operation and then generates more heat then is required, or it is off and generates no heat at all. As a result, the controlled condition is continually passing the desired value; the two seldom correspond for any length of time.

As the outside temperature measured by *OD* drops, the fluid in its phial contracts, as a result of which the contact operated by the bellows closes at a flow temperature which, in warmer weather, would have satisfied *FD*. Thus its control point tends to rise by an amount given by the dimensions of the two phials and their relative temperatures. Whether or not this rise is enough to offset the increased transmission losses is beyond the scope of the control system to determine, and to this extent its open-loop nature is confirmed.

Both processes are subject to *disturbances u*. In fact, it is in order to overcome the effect of such disturbances that control systems are installed. In the case of the space temperature controller, disturbances can consist of the opening or closing of doors or windows, changes in ambient or flow temperatures, the generation of heat in the space due to occupancy, solar radiation or otherwise. In as far as the heat gains and losses indicated by the letter q in Fig. 1.1 vary, they can be considered disturbances.

In the boiler temperature regulating system, the disturbances might consist of a change in the rate of flow, for instance, due to the opening up of more radiators, changes in heat requirements or, for that matter, changes in the calorific value of the fuel oil.

Straight line and wiring diagrams are used throughout industry. Block diagrams, on the other hand, of which Figs. 1.2 and 1.3 are examples, refer particularly to control circuits. They were developed to show the action of the components of the system and how they influence one another rather than how the individual elements are connected. Nor is the principle on which their action depends and by which the signals are transferred (i.e. electric, hydraulic, mechanical, and so on) relevant. On the other hand, although these diagrams are given here with the specific object of illustrating the nomenclature of automatic control and, by dotted lines, how the various elements may be combined to form the component parts of the control scheme, they are not typical. Some very simple examples of block diagrams as they are generally used and which may become highly intricate, will be found in later chapters of this book.

A small fraction of the 162 expressions defined in *B.S. 1523* and of the numerous alternatives in common use have now been reviewed. Although many of these are of use only to readers of more theoretical texts, a great number will be introduced in the following paragraphs where the concepts to which they refer are dealt with in more detail.

1.2. Controllers and control quality

Neither price nor sophistication of a controller provides a reliable yardstick for its suitability in a given case. An expensive electronic controller may 'hunt' in applications where a simple two-step controller would have given adequate results.

It will be shown that, from the control engineer's point of view, the characteristics of the process and those of the mode of the controller applied are of equal importance. The way these are matched will ultimately determine whether or not the results are satisfactory. Once the wrong type of controller has been installed, the cost and inconvenience of replacing it by a more suitable type may be considerable. Yet it may be preferable to face them rather than having to put up with an inadequate control system for its natural life of ten years or more. As prevention is better than cure, the first step must be to obtain a clear appreciation of the factors involved.

As far as the process is concerned, these characteristics will prove to be the same whatever variables are to be controlled. They can be summarised as the response time of the process to disturbances which the controller is intended to counteract. The response of the process is mainly determined by time lags of various kinds. These may have a favourable effect, but are usually detrimental to the quality of the control obtained. The second main feature is the *control range* or *span*,* *Sp*, of the process, as this determines the value of the largest potential deviation the controller will be called upon to neutralise.

It is true that at the design stage the response will seldom be predictable with any degree of accuracy. It should also be appreciated that such control 'constants', as they are referred to, may, in practice, be far from constant. Nevertheless a rough estimate will usually be sufficient to avoid the worst mistakes.

In this chapter the main control constants will be described systematically and in some detail. They are usually investigated by applying a disturbance of known character and magnitude and observing the way the process responds.

1.2.1. The disturbances

A disturbance, indicated by the symbol u, is any signal, other than the reference input, which tends to affect the controlled variable. In practice

*Referred to in the 1970 IHVE Guide as the controlled condition differential.

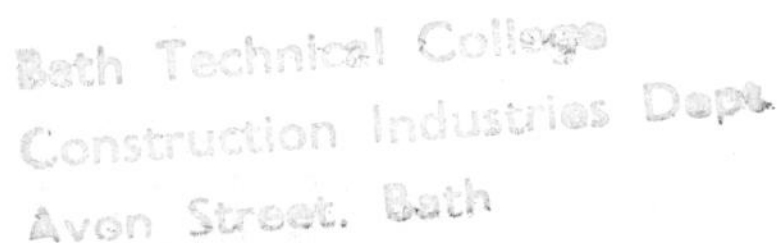

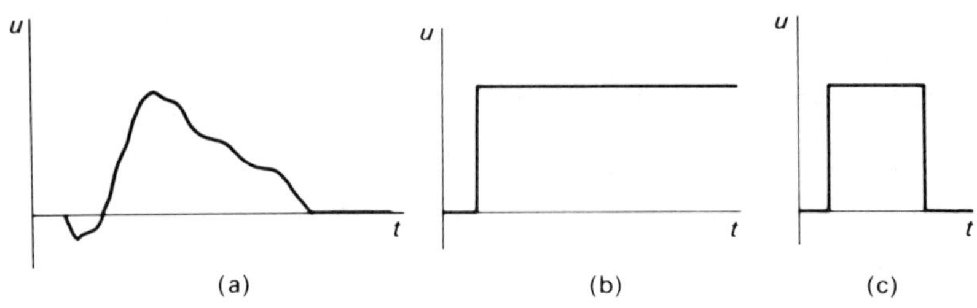

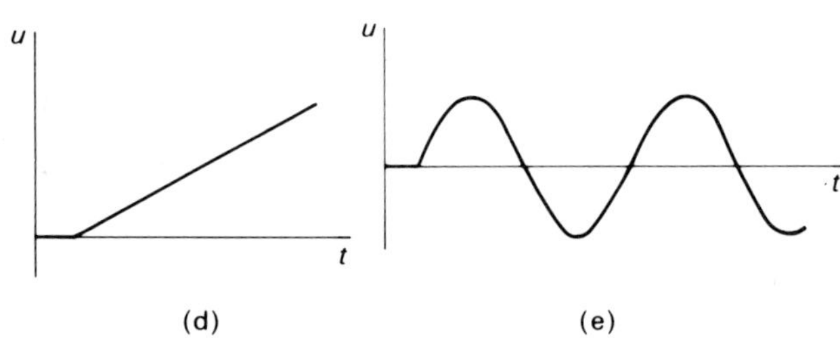

Fig. 1.5

Types of disturbance u.
(*a*) random; (*b*) single step; (*c*) double step; (*d*) ramp; (*e*) undamped sinusoidal.

most disturbances are random in nature, magnitude and frequency with which they occur, and they are therefore impracticable as a means of investigating the characteristics of the process to be controlled. For this purpose a number of clearly defined types of disturbance have been adopted (Fig. 1.5). The one most commonly used is undoubtedly the step change shown at (*b*). The idea is to see how the process reacts to a sudden change in load. Though this may be difficult to realise in practice it is easily and quite satisfactorily simulated by a sudden change in the desired value of γ_i. Sometimes two step changes of opposite sign are made in succession (*c*), spaced so as to allow the final control element to take up a significantly new position before returning to the one from which it started. The use of a ramp disturbance (*d*) is generally confined to laboratories etc. where the elaborate equipment necessary to produce it is available. The one and only type of disturbance that provides the control engineer with answers to all his questions is the cyclic disturbance (*e*), i.e. an undamped sine wave of adjustable frequency and amplitude. The reason is that all cyclic disturbances (*a*) can be resolved into a series of sinusoidal components. If the reaction of the controlled condition to each such component is found to be satisfactory, the compound signal will provide no difficulities. Although undoubtedly attractive to an engineer with a theoretical turn of mind, its practical use is largely restricted to processes reacting rapidly, such as the control of an electrical voltage or of a pressure in a closed vessel. As the reaction becomes slower, the equipment required to generate the necessary signals become both expensive and unwieldly. The experiments take too long and their cost is out of proportion to the results. To the practical engineer for whom this book is intended, therefore, this method of investigating a control circuit is likely to remain a theoretical consideration.

1.3. The 'constants' of the process being controlled

1.3.1. The response curve

Consider the 'process' of steam flowing through a short length of pipe in which the rate of flow, the 'controlled condition' is to be maintained at a constant value by means of a control valve. If the latter is opened or closed by a small amount, the rate of flow will be affected immediately and the new steady value will be reached practically at once (Fig. 1.6 (*a*)). As long as there are no disturbances, such as fluctuations in the supply pressure or changes in the resistance of the circuit, the new value will be maintained indefinitely.

A process of this kind is called *a process of zero order*. They are comparatively rare. A near approach, shown at (*b*) is the pressure in a confined space in its reaction to a change in damper position.

Generally, however, it will take longer for the controlled condition to reach its new steady state after a step change. Take the case of a thermometer dipped in boiling water. Initially, the mercury thread will rise quickly, slowing down more and more until it appears to come to a standstill. Yet, leaving the thermometer in position for a few minutes a further change, however small will be discernible. Theoretically, the ultimate position of the mercury thread will never be reached. For the mercury in the bulb to expand, heat must be transferred to it which is only possible as long as a temperature difference $\Delta\theta$ between heating medium and heated object provides the driving force. As time goes on, $\Delta\theta$ becomes smaller and smaller, and with it both the transfer of heat

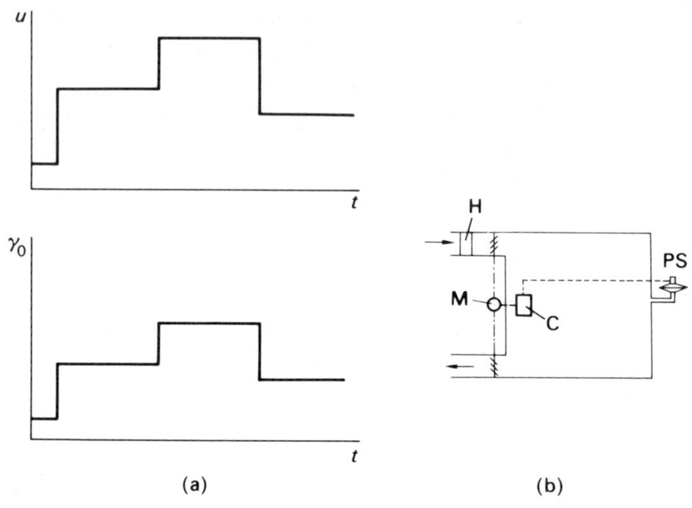

Fig. 1.6
Processes reacting instaneously to a disturbance; (*a*) the reaction of the controlled variable to a series of step changes; (*b*) a process with practically instantaneous reaction: the control of air pressure in a confined space.

and the rise in mercury temperature. Thus, it is not possible to determine at which point in time the thermometer exactly indicates the water temperature. It is instructive to plot the temperature indication against time as the curve thus obtained will prove to have a number of interesting characteristics. Assume that the initial temperature of the thermometer was 20°C and that the mercury thread takes 1s to reach the 60°C mark, i.e. to complete one half of the range ($\Delta = 100 - 20 = 80$°C). A further second will pass before one half of the remaining 40°C is completed, yet another second for the next 10°C and so on (Fig. 1.7). The steady reduction of the gap remaining between actual and ultimate values illustrates in a different way the fact remarked upon earlier, that, theoretically the two will never coincide. Similar curves would have been obtained had the step change been a multiple or a fraction of the 80°C used as an example (Fig. 1.8), or if the condition observed had not been temperature but any other variable such as a liquid level, the pressure in a large air tank, the relative humidity in a tobacco store, the voltage across a capacitor when charged through a series resistance.

The method of construction shows that the tail end of each curve

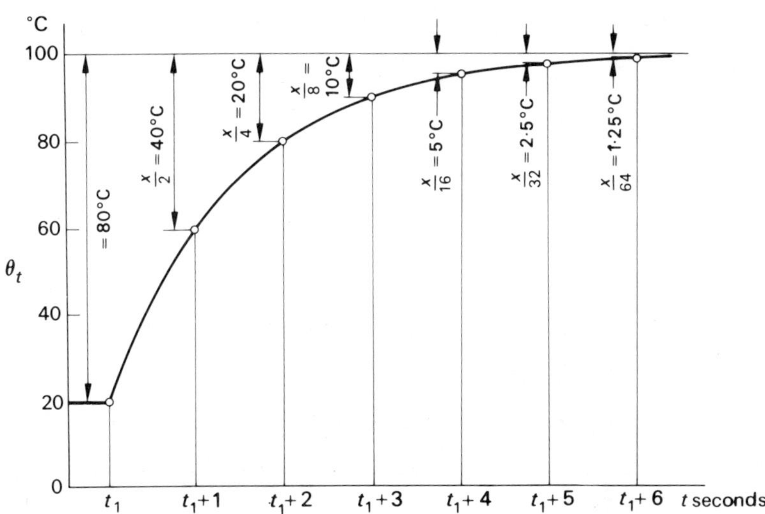

Fig. 1.7
Rate of change in temperature, θ_t, of a thermometer dipped into boiling water at a time t_1.

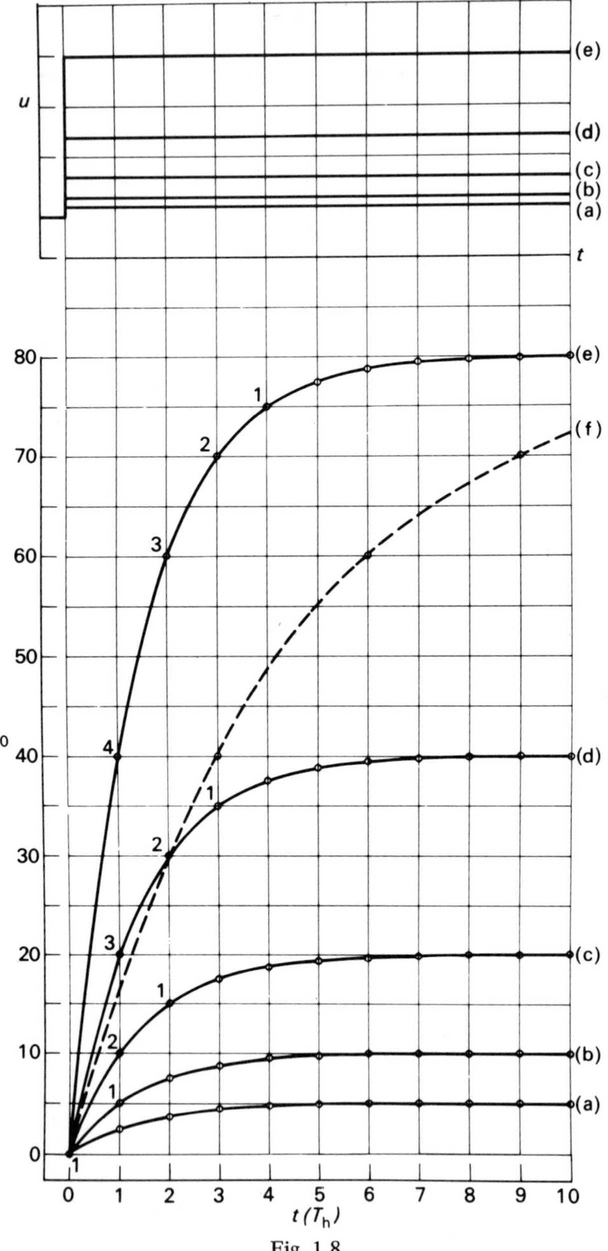

Fig. 1.8
(*a*) to (*e*) The reaction of a process similar to that of Fig. 1.7 to steps of varying magnitude; (*f*), a process reacting more slowly to step change (*e*) The half-life law applies in each case independent of the magnitude of the step change.

corresponds exactly to its predecessor. Thus the shape of (*a*) in Fig. 1.8 is identical with that of the others from point 1 onwards, that of (*b*) with that of the others from point 2 onwards, and so on. The figure also shows a dashed curve (*f*) of which the different appearance is wholly due to a slower reaction of the process to disturbance (*e*). It can be reduced to the earlier curve (*e*) by choosing a different time scale.

The rate of response is an important feature of any process to be controlled, and it would be useful to have a yardstick for purposes of comparison and definition. To this end, the time taken to *complete* the step was found to be useless (being infinite); that necessary for the first half of the ultimate step, on the other hand, is quite definite. For this reason physicists and geologists have adopted the so-called 'half-life period' as a measure of the speed with which processes, such as the spontaneous transmutation of unstable elements, react. This half-life time, T_H, was used as time scale in Fig. 1.8. Once this and the magnitude of the step is known, the curve is wholly defined. It is interesting to note that six half-lives must elapse before the ultimate value is approached to within 2 per cent.

Theoretical reasons have lead the control engineer to adopt a different unit with which to measure the rate of response. It is called the *time constant*, τ, of the process and is closely related to the half-life. It equals the time taken by a system to pass through 63·2 per cent of the ultimate change in controlled condition rather than the 50 per cent step of the physicist, and it is important enough to warrant a closer examination.

1.3.2. The time constant τ_1

Consider the well insulated storage calorifier of Fig. 1.9 containing Wkg of water at a temperature θ_1. It is heated by a coil kept at a mean temperature θ_2. If no water is withdrawn and heat losses are neglected, the calorifier temperature will rise slowly until $\theta_1 = \theta_2$, in accordance with the considerations detailed in the previous section. During the short time dt the calorifier will assimilate a quantity of heat

$$\mathrm{d}H = UA\,(\theta_2 - \theta_1)\,\mathrm{d}t \qquad (1.1)$$

due to which its temperature will rise by an amount dθ: if c is the specific heat capacity of water then:

$$\mathrm{d}H = cW\,\mathrm{d}\theta_1 \qquad (1.2)$$

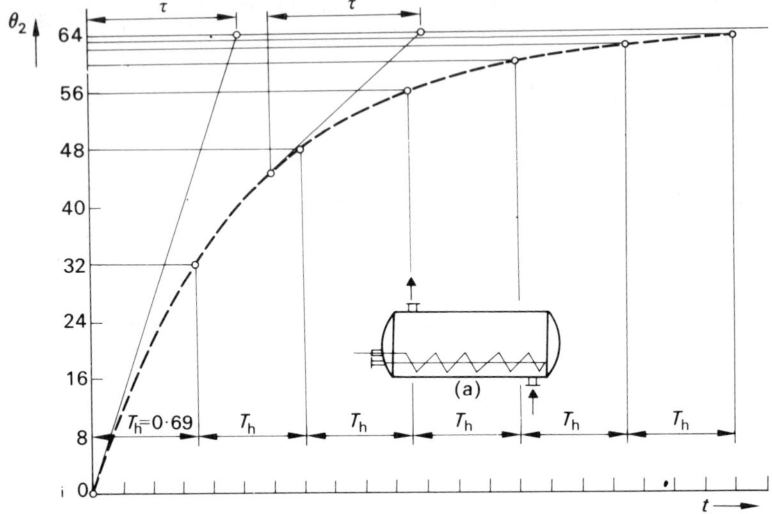

Fig. 1.9
The rate of reaction of a lagged calorifier. The figure shows a simple way of constructing an e-curve.

Equations 1.1 and 1.2 combined give

$$UA\,(\theta_2 - \theta_1)\,\mathrm{d}t = cW\,\mathrm{d}\theta_1$$

$$\frac{\mathrm{d}\theta_1}{\mathrm{d}t} = \frac{\theta_2 - \theta_1}{\dfrac{cW}{UA}} = (\theta_2 - \theta_1)\frac{1}{CR} \tag{1.3}$$

In this expression the heat capacity of the calorifier cW is assigned the symbol C, i.e. the heat capacity of the calorifier plus that of the water it contains, and the resistance to be overcome by the flow of heat i.e. the inverse of the transmittance $R = 1/UA$. Rearranging eqn 1.3:

$$\frac{\mathrm{d}t}{CR} = \frac{\mathrm{d}\theta_1}{\theta_2 - \theta_1} = -\frac{\mathrm{d}(\theta_2 - \theta_1)}{\theta_2 - \theta_1}$$

After integration one obtains:

$$-\frac{t}{CR} = \ln(\theta_2 - \theta_1) \text{ or } \theta_2 - \theta_1 = \exp(-t/CR) = \frac{1}{\exp t/CR} \tag{1.4}$$

a so-called e-function. At a time $t = CR$, $\theta_2 - \theta_1$ will be reduced to

$e^{-1} = 1/2{\cdot}718 = 0{\cdot}368$ times its original value. In other words, during this time the temperature has risen by $(1 - 0{\cdot}368)$ 100 per cent = 63·2 per cent of the step change.

As shown by eqn. 1.3, CR also denotes the time the process would have needed to reach the ultimate temperature θ_2 if the initial rate of heat transfer as indicated by the slope of the tangent $d\theta_1/dt$ had been maintained throughout. This is illustrated by the tangent in Fig. 1.9 and, in view of what follows it is a feature of considerable importance. Provided the relevant value of θ_1 is used, it does not matter at which point of the curve the tangent is applied: a τ of equal magnitude is always found.

It applies equally to any other variable operative in systems consisting of a single resistance and a single capacitance, provided C and R are expressed in the correct units (see Table 1.1). Though the definitions are clearly set out, real understanding of the conceptions C and R for different variables may initially prove difficult. Thus, when the controlled variable is a temperature, C is the amount of heat that has to flow to the process if the temperature is to rise by one degree; R is the temperature difference required to transfer 1J per unit of time to it. In the case of a liquid level to be controlled, C is the amount of fluid to be supplied to a vessel in order to raise the level by one unit of length; R is the pressure difference required, expressed in height in meters of a water column to cause one unit of volume per unit of time to flow into the container.

Table 1.1

Variable	C		R		$\tau = CR$
Temperature, θ	$\Delta Q/\Delta\theta$	J/°C	$t\Delta\theta/\Delta Q$	h°C/J	h
Pressure, p	$\Delta V/\Delta p$	m^3/bar	$t\Delta p/\Delta V$	h bar/m^3	h
Liquid level, l	$\Delta V/\Delta h$	m^3/m	$t\Delta h/\Delta V$	h m/m^3	h
Rel. Humidity, ϕ	$\Delta W/\Delta\phi$	kg water per % rel. humidity	$t\Delta\phi/\Delta W$	h%rH/kg H_2O	h
Electric charge, q	$\Delta q/\Delta V$	coulomb/volt	$t\Delta V/\Delta q$	h volts/coul.	h

1.3.3. The single-stage response curve

After a time equalling 2τ, the ultimate value will be approached to within $(1/2{\cdot}718)^2 \times 100$ per cent = 13·6 per cent; after 3τ to within $(1/2{\cdot}718)^3$ or 5 per cent, after 4τ to less than 1·8 per cent. In 1.3.1 it was mentioned that $6T_h$ must elapse if the ultimate value is to be approached to within $(\frac{1}{2})^6 \times 100$ per cent or 1·56 per cent. The ratio τ/T_h thus approximates to 4/6. The exact ratio proves to be 0·69 or,

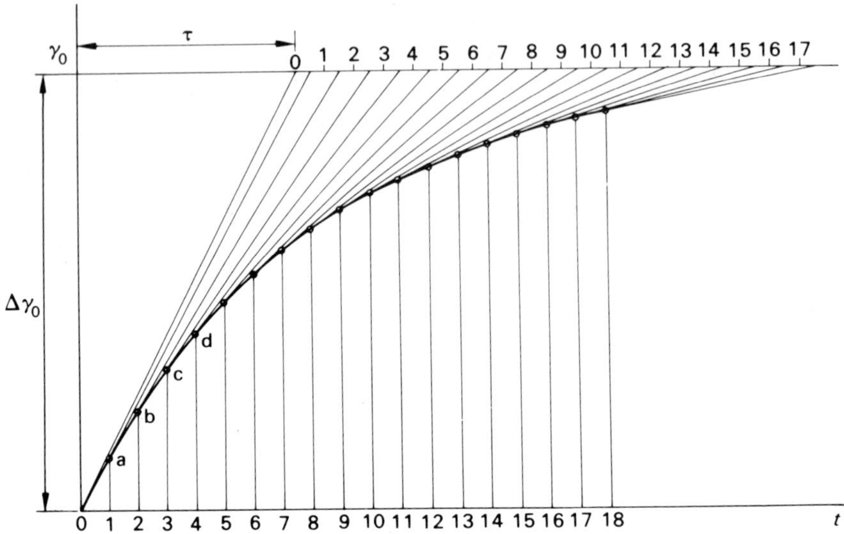

Fig. 1.10
A more accurate approximation to an e-curve.

roughly, 0·7. With this in mind it is possible to draw an exponential-curve with fair accuracy, by plotting the initial and ultimate values along the ordinate and dividing the distance between the two into multiples of 2 units, e.g. 32 or 64. The abscissa is divided into units of which, if τ equals 10, T_h will equal 7. Now the tangent at the foot of the curve can be drawn (Fig. 1.9), as well as the points $(T_h, 32)$, $(2T_h, 48)$, $(3T_h, 56)$ and so on, the tangent at the foot of the curve helping to bridge the gap between the points (0,0) and $(T_h, 32)$.

A more accurate approximation is shown in Fig. 1.10. Here the abscissa is scaled with divisions as near to one another as accuracy requires. The asymptote is found as a line parallel to the abscissa at a distance equal to the ultimate step $\Delta\gamma_0$; along this line the time constant τ is plotted. Starting from point $(\tau, \Delta\gamma_0)$, repeat the scale of the abscissa. The tangent 0–0 shows the initial slope of the curve. During the period 0–1, the mean rate of rise will be such that the curve will strive towards point $\frac{1}{2}$ on the asymptote and the intersection *a* between line 0–$\frac{1}{2}$ and ordinate 1 will lie on the curve with greater accuracy the closer the horizontal scale divisions lie together. Point *b* of the curve is

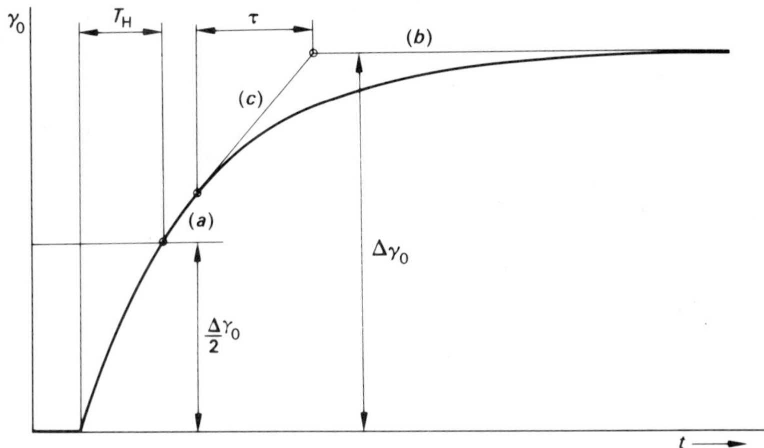

Fig. 1.11
Determining τ and T_H from an experimental curve (*a*); (*b*) is the asymptote; (*c*) is a tangent drawn at an arbitrary point, A, on the curve, preferably situated where the rate of change is maximum.

found on the intersection of line $a - 1\frac{1}{2}$ and ordinate 2, similarly, point c on the intersection of $b - 2\frac{1}{2}$ with ordinate 3, and so on.

If, on the other hand, the response curve of a single *RC* unit is obtained from measurements, both T_h and τ are readily determined by graphical methods (Fig. 1.11).

Should, in time, the heating coil be fouled by internal or external deposits, *R* will increase. Thus it is clear that the main 'constant' of the process, τ, will also increase, and it is, therefore, by no means constant. In thermal applications τ may also be affected by the magnitude of the step change, since the larger it is, the more vigorous the convection currents and the greater the factor *UA* is likely to be. Thus, the constancy of such a 'constant' should be viewed critically. In this respect, determining τ experimentally has the advantage over the mathematical method; on the other hand, the latter has its undoubted advantages due to the difficulty of drawing asymptote and tangent with any accuracy, especially as the curve will seldom approach the ideal shape of Fig. 1.11. As a check, therefore, calculation has a useful function. Here the reader should bear in mind the definition of τ given in Section 1.3.2, namely that τ equals the time needed by the controlled variable, when subjected to a step change, to reach its ultimate value if the initial rate of change

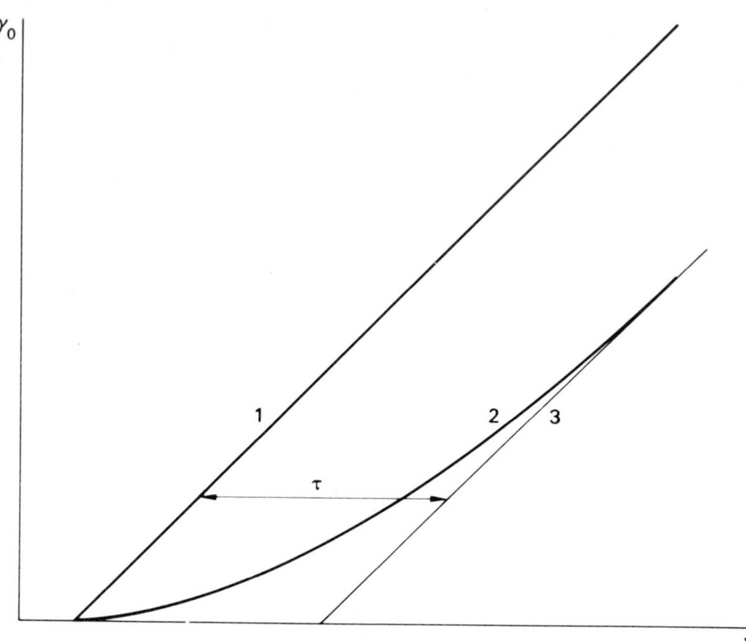

Fig. 1.12
Response to a ramp disturbance e.g. 1, ambient temperature; 2, process temperature; 3, asymptote.

were maintained. Let the calorifier of Fig. 1.9 weigh 450 kg when empty, and hold $2{\cdot}5\ m^3$ of water, then the factor $C = cG$ equals

$$C = (2500 \times 1 + 450 \times 0{\cdot}114)\,4{\cdot}19 = 10{\cdot}7 \times 10^3\ \text{kJ/°C}$$

If, at a mean temperature difference of 50°C, the coil produces 1500×10^3 kJ/h,

$$R = \frac{50}{1500 \times 10^3} = 0{\cdot}033 \times 10^{-3}\ \text{h°C/kJ}$$

Thus,

$$\tau = CR = 0{\cdot}033 \times 10^{-3} \times 10{\cdot}7 \times 10^3 = 0{\cdot}4\text{h approximately}$$

It will be noted, that in this calculation heat losses due to connection and radiation have been neglected. In heating and air conditioning, very small values of τ are also encountered. Thus, in water or air mixing processes, τ may well be of the order of a second or less.

So far, the response to a step change has been used to elucidate the concept of the time constant, as this is the way in which it is usually determined in practice. There is, however, an alternative method used in laboratories in connection with the special equipment that is required, namely the response to a ramp change (Fig. 1.12). It is found, that, in this case, the controlled variable will gradually approach an asymptote parallel to, and at a distance τ from, the original ramp change. In practical terms, Fig. 1.12 could illustrate how, starting from a balanced condition, the temperature of an object changes (line 2) when exposed to a linear rise in ambient temperature. The vertical distance between the two heavy lines denotes the dynamic error introduced by the fact that *CR* is finite.

1.3.4. The response to heat extraction

From a control engineer's point of view, the difference between the flow of energy to a process or its extraction from it is only a question of sign (Fig. 1.13). Let us assume that, in cooling down, the temperature

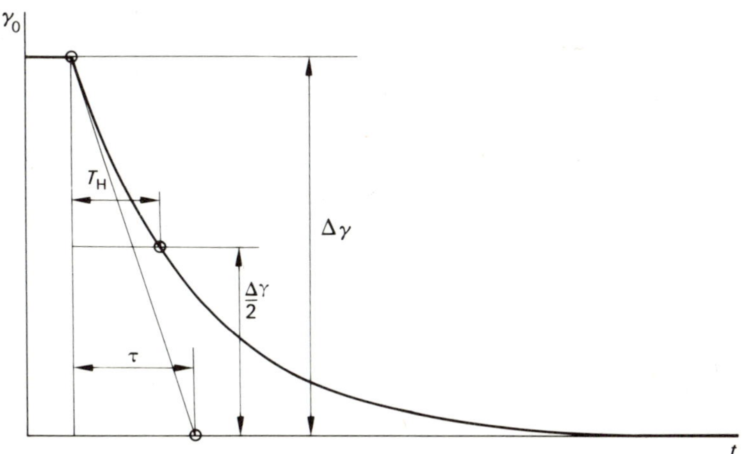

Fig. 1.13
Response of a single *RC* unit to a step change with negative sign.

distribution in the calorifier described in the previous section remains homogeneous (which is, in fact, seldom the case). Then, when hot water is withdrawn and replaced by an equal volume of cold water, the temperature will initially, drop quickly. Once more it approaches the ultimate temperature asymptotically, slowing down as θ is reduced. In this case, also, the process responds in accordance with an e-curve, characterised by the product *CR*.

C is not normally affected by the direction of the heat flow; R, on the other hand, may assume a very different value. Whereas in any heat exchanger the transfer of heat to the heated medium is assisted by providing a large heating surface along which the circulation is unimpeded and the turbulence of the heated medium is, as far as possible, ensured, the cooling down process is retarded for economical reasons, e.g. by an efficient lagging. Thus, in heating installations, the time constant of the heating cycle is usually shorter than that of the cooling cycle; in cooling installations the reverse applies.

Supposing the calorifier of the preceding section is insulated in such a way that the heat losses due to convection and radiation equal 5% of full capacity, then, if no hot water is drawn off and at an ambient temperature of 15°C, the resistence to discharge will equal

$$R_d = \frac{100}{5} \times 0{\cdot}033 \times 10^{-3} = 0{\cdot}66 \times 10^{-3}\ \mathrm{h \cdot C/kJ}$$

and the cooling down curve will be characterized by a time constant

$$\tau_d = CR_d = 10{\cdot}7 \times 10^3 \times 0{\cdot}66 \times 10^{-3}\ \mathrm{h} = 7\mathrm{h}\ \text{approximately.}$$

When hot water is withdrawn, the picture changes completely. Assume that the temperature has risen to 55°C as a result of which the control valve in the flow has closed and remains closed even though withdrawal takes place. If (which is very unlikely) the cold water flowing into the calorifier at 15°C mixes instantaneously and completely with the remaining hot water which is withdrawn at a rate of 75 l/min or 4·5 m^3/h, heat is initially withdrawn at a rate

$$4{\cdot}5 \times (55 - 15) = 180 \times 10^3\ \mathrm{kJ/h}$$

then,

$$R = \frac{40}{180 \times 10^3}\ \mathrm{{}^\circ Ch/kJ}$$

and as C is unchanged,

$$\tau = RC = \frac{40}{180 \times 10^3} \times 12{\cdot}1 \times 10^3 = 2{\cdot}7\ \text{hours}$$

1.3.5. The time constant in processes where charge and discharge are simultaneous

This is the usual condition: as soon as the temperature of a boiler, calorifier or room starts to rise above ambient a certain loss of heat will take place. In order to visualise what then happens it may be convenient

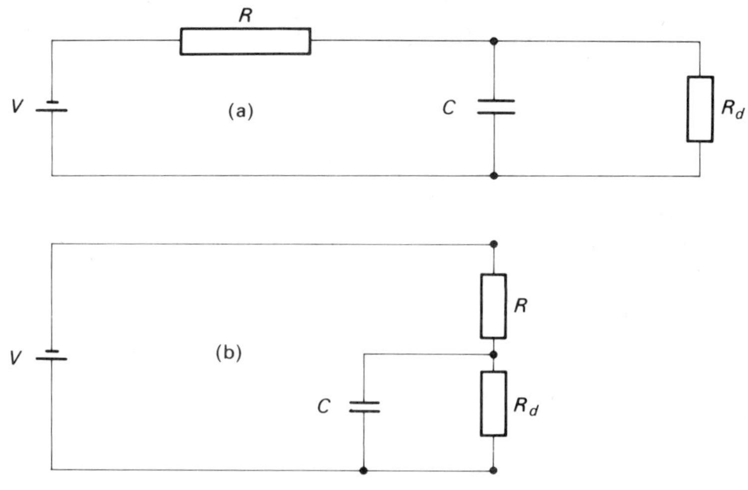

Fig. 1.14
(*a*) A capacitor is charged by a battery of voltage V through a resistance R and bypassed by a further resistance R_d; (*b*) the same circuit rearranged.

to use a procedure widely applied in connection with analogue computers, i.e. to simulate the process by an electric circuit. Figure 1.14 shows a capacitor C charged by a battery V through a resistance R while the charge is continuously drained off by a further resistance R_d. By rearranging the circuit (*b*) it is obvious that the capacitor charge will tend to reach a maximum voltage equalling

$$V_c = \frac{R_d}{R + R_d} \times V$$

As is shown in Fig. 1.15 the rate of change at the foot of the curve is not affected; some charge must have accumulated in the capacitor before discharge can occur. Therefore the slope of the tangent at the foot of the curve remains the same and, as simple geometry shows, τ is reduced in the same ratio as the span, Sp. If $R_d = R$,

$$\tau' = \frac{R_d}{R + R_d} \times \tau = \frac{\tau}{2} \text{ and } Sp' = \frac{Sp}{2}$$

Any heating engineer will confirm from his own experience that as a boiler is overloaded, both the flow and the return temperatures will drop. At the same time, R is reduced as a result of the increased load and with it τ.

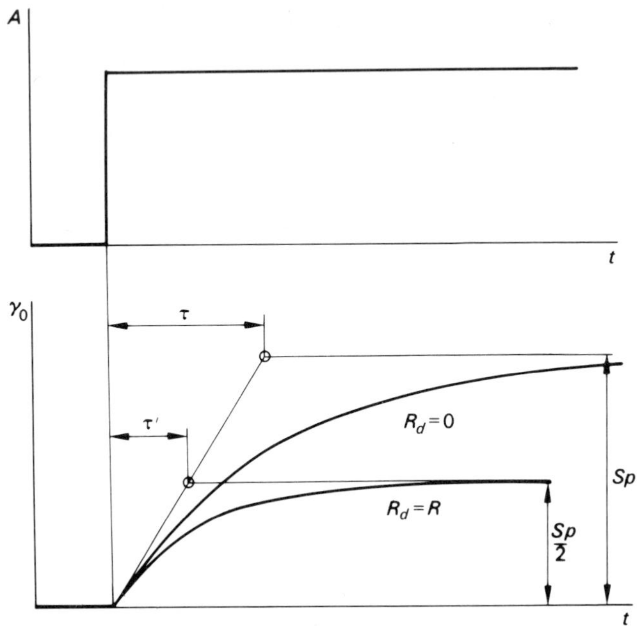

Fig. 1.15
The slope of the tangent at the foot of the curve is not affected by the presence of a finite R_d, but both τ and Sp are reduced.

In Fig. 1.16 response curves are shown for different initial conditions. Curve 1 is the reaction to charge at infinite discharge resistance. Once the maximum value has been reached, the charge is discontinued and the discharging process is started (curve 2; $R_d = 2R$); curve 3 refers to a simultaneous charge and discharge $R_d = 3R$ at (*a*) and $R_d = R$ at (*b*). If the charging process had started under load, the response would have been in accordance with curve 4. Whatever the initial conditions, the ultimate condition is the same for a given ratio of R to R_d. This should not be taken to mean that curves 3 and 4 at (*b*) are necessarily symmetrical.

The influence of load on τ and Sp is of considerable importance. Take the reaction of an ordinary boiler thermostat controlling a flow temperature, θ_F, by switching the burner on or off to the change in θ_F it has initiated itself. If the stem is surrounded by hot water over the whole of its length, the value for A will be maximum, and for R a mini-

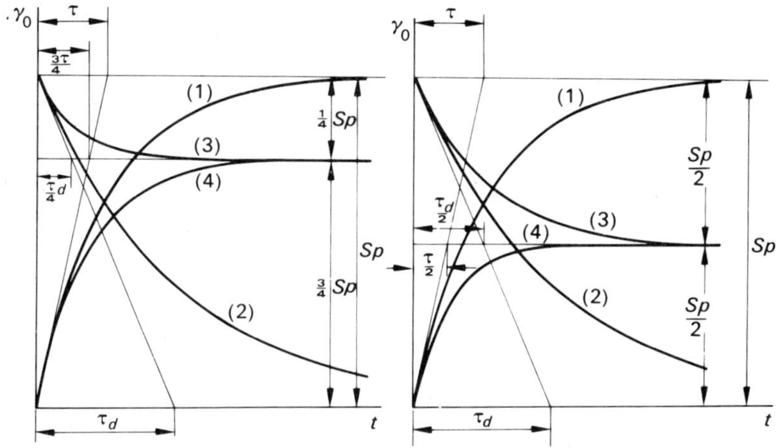

Fig. 1.16
Response curves shown for various initial conditions and ratios R_d/R

mum. As, furthermore, the *C* for a thermostat is relatively small, the reaction to a change in temperature will be reasonably quick. If on the other hand, the fit of the stem in its pocket is poor and the latter is only partly immersed in the water stream, *R* will increase and the response will become more sluggish. R_d will increase and *Sp* will drop; the flow temperature will have to reach a higher value before the thermostat switches, the instrument will show calibration errors which increase if its head is subjected to draught (i.e. if R_d decreases). The case of a room with on/off control is more complicated still. During the charge phase, both the *UA* of the radiators, which is relatively large, and the smaller *UA* of the walls (variable in itself with wind velocity and direction) are operative; during the discharge period it is the latter only. The *C* varies too; opening the valve causes immediate ejection of the cool radiator contents.The *cG* of the radiator is negligible. Yet, on closure, the *cG* of the radiator is added to the *C* of the room. As a result of these and other factors τ_d is frequently considerably larger than τ, and the time constant for simultaneous charge and discharge is smaller than either τ or τ_d.

It need not be emphasised that similar considerations may apply *mutatis mutandis* to whatever the controlled condition may be. The τ

and the *Sp*, of the water level in the vessel of Fig. 1.27(*a*) and of the boiler temperature at (*b*), are closely related to the rate at which the water is drawn off through the exhaust valve.

Again, in the stores of a printing plant in which relative humidity ϕ is to be controlled, the reaction to a constant supply of steam will be more rapid, and the ultimate level of ϕ will be higher when the stores are empty than when a new supply of relatively dry paper has just been received.

Comparing the response curve of Fig. 1.6, a system in which $RC = 0$, with the subsequent response curves, it will be seen that an *RC*–stage introduces a time lag; the reaction of a process is retarded by the dynamic error referred to in Section 1.3.3. This effect, especially apparent in Fig. 1.12, is therefore called an *exponential time-lag*.

In contrast with other types of lag, all of which make the controllability of a process in which they occur more difficult, an exponential lag usually facilitates the control problem. The larger the time constant of the process, the slower will the controlled conditon react to a given disturbance and the smaller will be the deviation during the time the detecting element needs to detect such a change; in addition, the control action can be more vigorous without causing objectionable overshoot. Its presence in a process allows the controller time to adjust itself to changes in the controlled condition without large deviations occurring in the meantime. This should emphatically not be taken to mean that a large *RC* element in the controller (i.e. in the detecting or the final control elements) is necessarily desirable, a question which will be returned to later on.

1.3.6. Two or more *RC* elements in series

Usually there are a number of *RC* elements, large and small, in series, as a result of which the response curve of the first order, i.e. for a single so-called exponential transfer stage, becomes distorted to that of a two-, three- or multi-stage system or of second, third or higher order.

The nature of the distortion depends upon the number and magnitude of the individual *RC* elements placed in series, and infinite variations are possible, although the general picture remains the same. It will prove helpful to study the simple example of Fig. 1.17, referring to two *RC* units, a boiler heating storage calorifier. The construction of Fig. 1.18 gives a fairly accurate idea of the way in which the calorifier temperature γ_c will rise if both units start from cold.

The two time constants τ_b and τ_c must be known and, in this case, are supposed to be equal. The response curves θ are drawn up on an

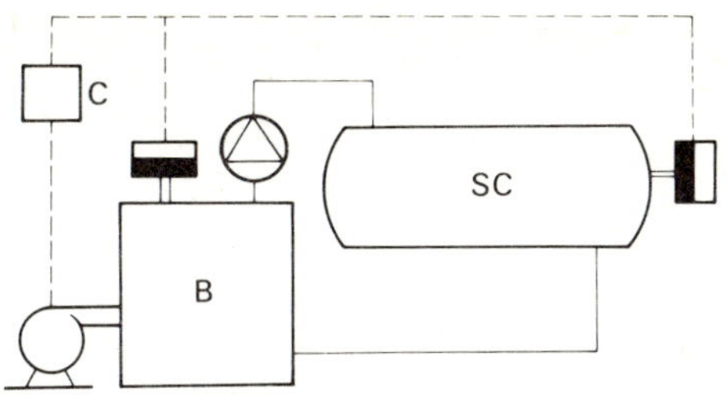

Fig. 1.17
Two *RC* units, a boiler, *B*, and a storage calorifier, *SC*, in series

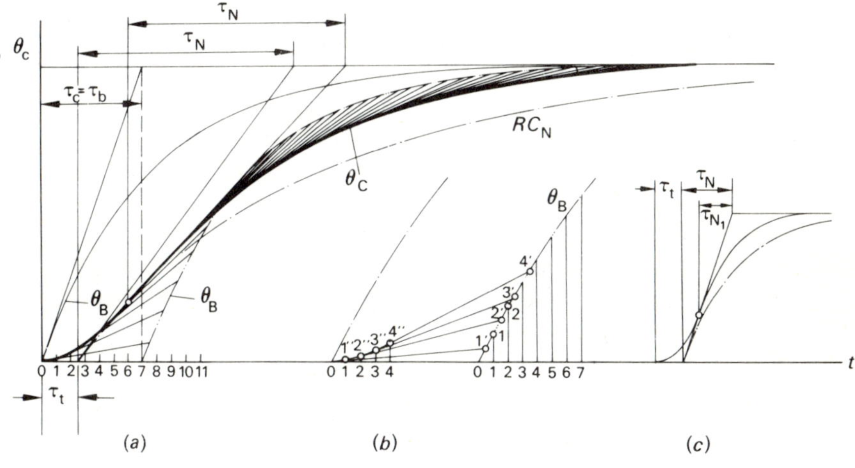

Fig. 1.18
Construction of the response to a step change in a two-stage system. γ_b and γ_c are the boiler and calorifier temperatures respectively

arbitrary time scale, the first one, in this case that of the boiler, θ_B, being shifted forward along the time scale by an amount τ_c, for the purposes of this construction. Referring to detail (*b*), θ_b will have reached point 1 after one unit of time has elapsed. Its average rate of rise during this period, however, would have corresponded to point 1′, and it is to this point that θ_c will strive. Point 1 inch on the intersection of the tangent and ordinate 1 will therefore lie on θ_c. During the second period, θ_b reaches 2, though the average rate only corresponds to 2′. Drawing the second tangent 1″– 2′ results in point 2″, etc. Thus the calorifier response curve slowly emerges as a series of short lines 0 – 1″, 1″ – 2″, 2″ – 3″ etc., as shown in Fig. 1.18 (*a*). If unequal time constants had been used, the result would have been different, but independent of the order in which they appear. The smaller values of τ tend mainly to affect the shape at the foot, the larger ones reduce the rate of change of the rest of the curve. Figure 1.19 shows how adding further (equal) *RC* units in series affects the response curve. In Fig. 1.19(*a*) the time scale is independent of n, the number of *RC* units: at (*b*) the scale divisions to which the *nth* order curve is drawn equal $1/n$ of those of the original *RC* curve. At (*c*) the effect is shown of dividing a single *RC* unit into n equal parts.

1.3.7. The notional time constant in multi-stage systems

The construction of such curves, although most illuminating, becomes a laborious and complicated matter. It is possible to express a curve of the second order mathematically, more difficult to do this for one of the third order, and beyond that it is no longer feasible. As, in most processes, the number of *RC* units, large and small is likely to be large, a simple approximation of such curves is generally used. The method is shown in Fig. 1.18(*c*), in which θ_c was found as a curve of the second order. At the osculation point a tangent is drawn. The time taken for the variable to reach its ultimate value if the rate of change at the osculation point were maintained throughout is now assumed to be the new time constant, τ_{1+2}, or the notional time constant τ_{N_1}. The accuracy, or rather the lack of it, is shown by the dash-dotted line at (*a*) which is a true expotential curve drawn to the tangent at the osculation point. It should be noted that the method by which the time constant was determined is not the one most commonly used. Usually the tangent is protruded in both directions and τ_N defined as the difference in time between the points of intersection with the asymptote and the abscissa respectively (τ_N). Thus τ_N and the discrepancies between the true and the approximated curve may become larger still (Fig. 1.18(*c*)).

In practice, response curves for multi-stage systems are obtained experimentally. This usually means (Fig. 1.20) that they do not run as smoothly as the curves of Figs. 1.18 and 1.19, so that the angle the tangent makes with the abscissa, as well as the position of the asymptote denoting the ultimate value, become uncertain. As the value of τ_{res} is highly dependent upon both, it is clear that one should not rely too heavily on the results obtained by graphical treatment of experimental data.

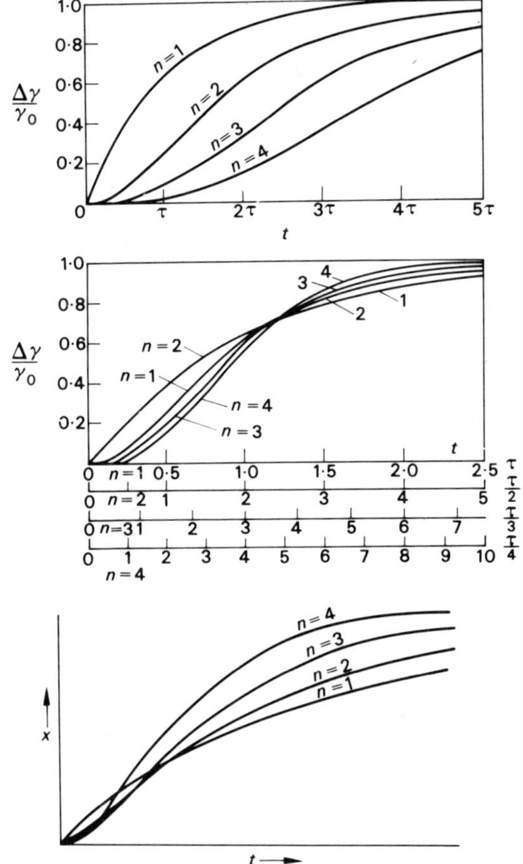

Fig. 1.19
Response curves of multi-stage systems.
(*a*) The effect of adding *RC* units of equal magnitude; (*b*) Part (*a*) shown on time scales inversely proportional to the number, *n*, of units; (*c*) the effect of dividing an *RC* unit into–equal parts and placing these in series.

1.4. Further lags

1.4.1. Transfer lag

In Fig. 1.18 the part of the abscissa cut off by the tangent through the osculation point is indicated by the symbol T_t, standing for transfer lag. This is the lag caused by interposing one or more *RC* units between the

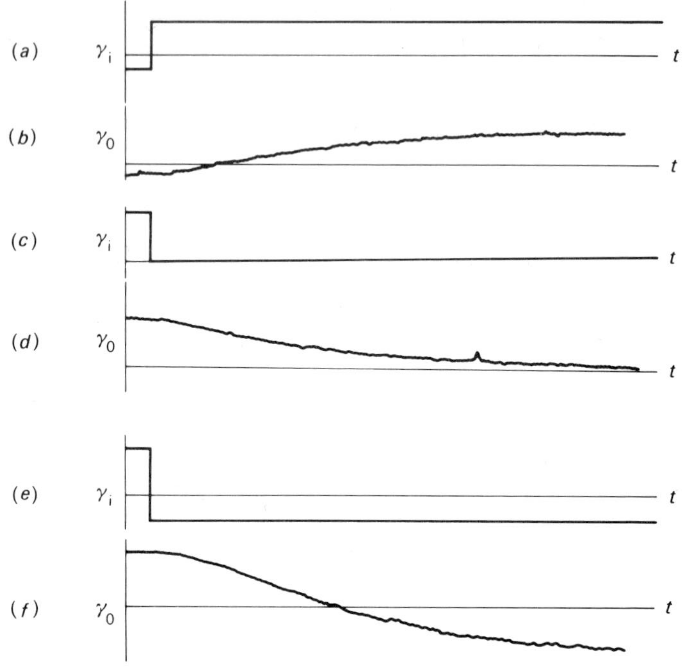

Fig. 1.20
Experimental step-change response curves of an air heater as measured by a sensing element located after a cooler battery and the fan. (*a*), (*c*) and (*e*) changes in set value; (*b*), (*d*) and (*f*) response curves.

source of, for example, heat and the process to be heated. The presence of transfer lag is clearly indicated by the distortion at the foot of the response curve. It means that some of the heat developed in the system in response to a need for it is stored on its way to the process. Though, ultimately, it will be liberated and will affect the controlled variable, its presence, for the time being, cannot readily be detected.

Thus it is impossible to draw valid conclusions as to when the supply has exactly met the demand. True, Fig. 1.18 showed that the presence of the boiler tended to slow down the rate of reaction of the calorifier, a circumstance usually beneficial to the 'controllability' of the process, but this small advantage seldom offsets the disadvantage of the transfer lag it introduces.

Take as an example space-heating control system in which a room thermostat is used to keep the temperature constant. If θ_c is taken to

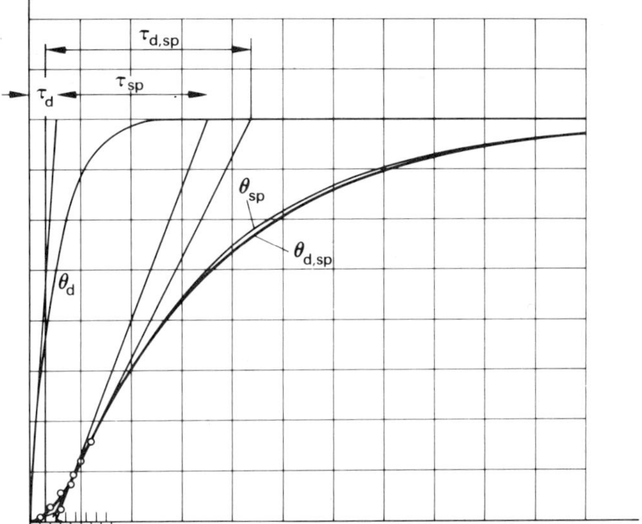

Fig. 1.21
Step function response of a heating plant as measured by a sensing element the time constant of which equals one fifth of that of the heating battery.

indicate the detector temperature and θ_b that of the space, Fig. 1.18 once more shows what happens, albeit in an exaggerated form. Assume that, on reaching point 4″, the thermostat is satisfied and closes its valve. In that time the space temperature, having reached point 4, will have changed nearly four times as much. Compare this with the arrangement of Fig. 1.21, in which the time constant of the sensing element τ_d has been reduced to approximately one sixth of its original value. When now the thermostat is satisfied, the space temperature will scarcely have changed one half as much as it had in Fig. 1.18.

This example shows how important it is to keep the time constant of a sensing element to as low a value as possible: the effect of placing an *RC* unit, however small, between process and comparing elements is out of all proportion to the magnitude of its τ. It is in this fact that the value of n.t.c. and wire detectors with their very short time constants lies. The illustration confirms that the effect on the shape of the foot is unmistakeable, whereas the main part of the response curve is hardly affected. The response of the sensing element to a ramp change illustrates the issue in a different way (Fig. 1.22). At (*a*) the reaction of two

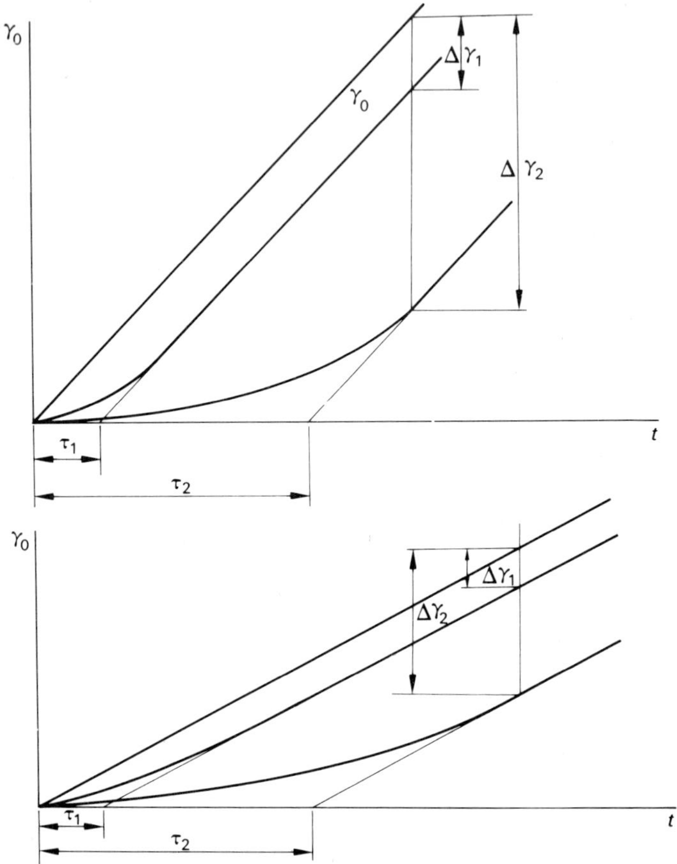

Fig. 1.22
The difference in lag in terms of the controlled variable γ caused by *RC* units of different magnitude when subjected to ramp changes with different rates of change.

detectors of widely different time constants, τ_1 and τ_2, is given; the lags $\Delta\gamma$, in terms of the controlled variable, are correspondingly different. When the rate of change is decreased (*b*), the ratio $\Delta\gamma_1/\Delta\gamma_2$ is unchanged but the absolute values will be less. This fact has an important bearing on the results to be expected in a given case. A sluggish sensing element may give acceptable results as long as the rate of change of the controlled variable is small; as it changes more rapidly, this result can only be maintained to the extent that the time constant of the sensing element can be kept down.

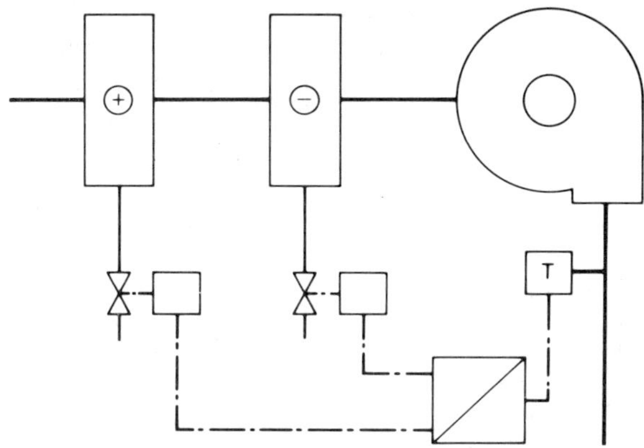

Fig. 1.23
The introduction of transfer lag by a cooler battery, interposed between the heating battery controlled and the detecting element, maybe a serious matter.

A further important conclusion may be drawn from the theory set out in these sections. Though the order of the *RC* units does not affect the shape of the ultimate curve, the way in which the controlled variable fluctuates in the various *RC* units does. Take the case of an air heater controlled by a very sluggish thermostat. The thermostat temperature will change but slowly, but the air temperature, i.e. the quantity it is required to keep constant, may fluctuate unacceptably. If a sensing element was chosen for its quick reaction without affecting the notional time constant of the combination, the dynamic error would be materially reduced. Thus the widely used (and very effective) method of reducing a tendency to hunt by the application of a sluggish sensing element or a slow-moving valve-motor is permissible only when the controlled variable changes slowly. Fortunately, this is frequently the case in heating and air-conditioning applications.

A different example is given in Fig. 1.23 where a preheater battery is controlled by a thermostat placed after the fan and therefore also after the cooling battery. During the heating season the water in the chiller is stagnant. Before the thermostat can detect the rise in temperature it has initiated itself by opening the heater valve further, the cooler must be warmed up. The delay thus introduced may seriously affect the control quality obtained.

1.4.2. Distance-velocity lag; dead time

This type of lag introduces a delay between the moment a disturbance occurs and that at which the process reacts, without the character or the magnitude of the response being affected in any way (Fig. 1.24).

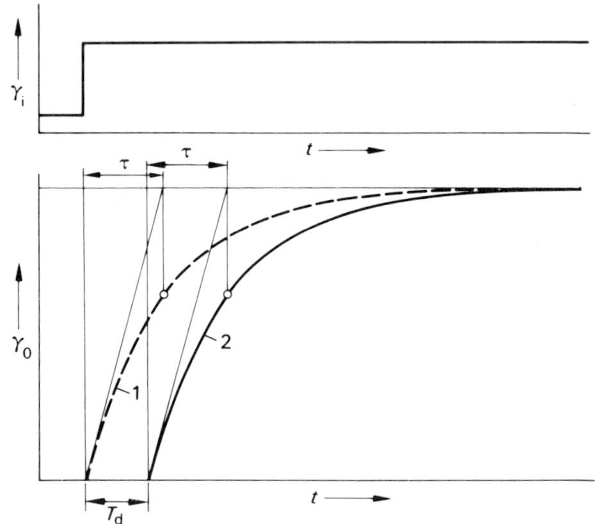

Fig. 1.24
Distance-velocity lag. The response to a disturbance is delayed by a time T_d but is otherwise unchanged.

A frequently used example is that of a flow detector located at a distance of 10m from a mixing valve, the position of which is suddenly changed. If the water velocity is 0·5m/s, it will take 20s for the new temperature front to arrive at the detector; the distance-velocity lag, T_d, equals 20s.

In actual fact, some heat will be transferred from the water to the walls of the pipe through which it flows. The result is the sharp step change measurable immediately after the valve is smoothed down. Therefore, the example is not truly representative of distance-velocity lag, which, in its true form, is rare. A better example, though neither distance nor velocity is apparent, is provided by an oil burner relay, at one time widely used but now obsolete. This introduced a pre-ignition time of 4 minutes after a call for heat from the controlling thermostat before the burner could come into operation. Once it started up, the

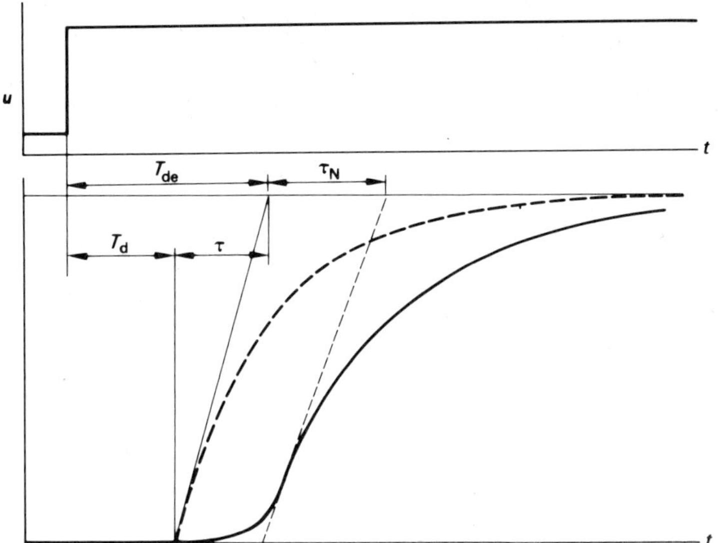

Fig. 1.25
Distance-velocity lag, T_d, transfer lag T_t, effective dead time T_{de}, notional time constants τ_N caused by placing a further *RC* unit in series with a unit with time constant τ.

boiler temperature would respond exactly as if no pre-ignition time had been interposed.

Inasmuch as *dead time* is defined as the time interval between a change in a signal and the initiation of a perceptible response to that change, distance-velocity lag is pure dead time. As long as the distance-velocity lag has not elapsed the controller is neither in a position to counteract the effect of a disturbance nor to correct that of any action it may have initiated. Since τ remains unchanged, it is more detrimental to control quality than transfer lag, but in as much as the storage effect is absent, it is less.

1.4.3. Effective dead time, T_{de}

In transfer lag proper, the dead-time element, as defined in Section 1.3.6, is negligible. Yet, in order to emphasise the essential difference between transfer and exponential lags, the term *effective dead time* is use to denote 'the time interval between the change in a signal to an element or system and the build up of the response to a specified proportion, say, until 5 per cent of the final change has taken place' (B.S.

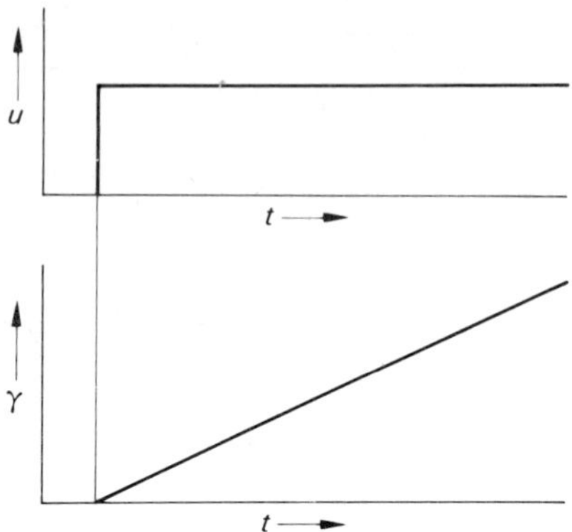

Fig. 1.26
Response to a sustained disturbance, u, of a process without self-regulation.

1523). In what follows, T_{de} will be determined as in Fig. 1.25 and defined as the sum of distance-velocity and transfer lags.

The adverse effects of transfer and distance-velocity lags, already mentioned, will be returned to later.

1.5. Self-regulation

Many simple processes are characterised by the fact that a new steady state is reached automatically, i.e. without interference by manual or automatic control. Thus, all processes to which a time constant can be assigned are inherently self-regulating to a greater or lesser degree. This property is usually valuable inasmuch as the magnitude of the largest deviation the controller will ever be called upon to correct is limited. Thus, if the heat supply to a room is increased, the temperature will not rise indefinitely as its rate of change is limited and progressively reduced by the increase in heat loss; in the course of time a new equilibrium is reached at a higher temperature.

Some processes not obviously characterised by an R and a C can also be self-regulating. The circulating head in gravity hot water systems shows similar effects. If the flow temperature is raised, the head will also rise owing to the increased difference between flow and return temperatures. The water flowing through the pipes and radiators will accelerate and reach a new equilibrium at a higher velocity, v, limited by the

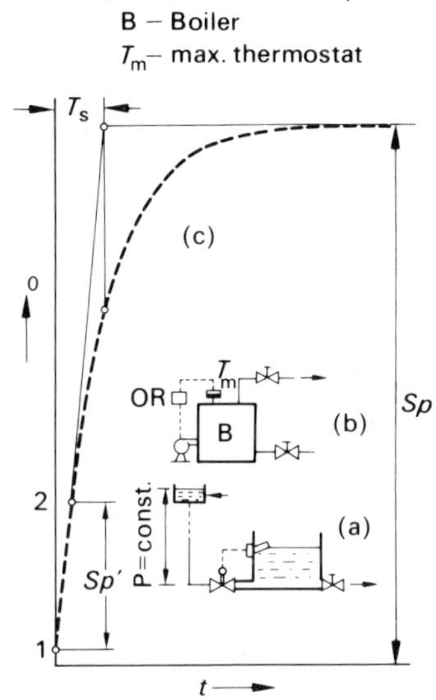

Fig. 1.27
(*a*) and (*b*) Two processes without effective self-regulation; (*c*) the control range, *Sp*, is large with respect to the range of values, *Sp*′, encountered in practice.

friction which rises with v^2. Then a second self-regulating factor makes itself felt. The water flowing more quickly through the radiators cools down less, the return temperature rises, the difference between flow and return temperatures is reduced and, with it, the gravity head; the water velocity settles down at a value lower than the value it reached soon after the increase in flow temperature occurred. It will be clear that a simple exponential curve, will not necessarily illustrate a process as complicated as the one described.

1.5.1. Processes with negligible self-regulation

Figure 1.26 shows the response of a process wholly without self-regulation. There are some processes in which, within the range of values of interest, the self-regulating effect, though present, is negligible (Figs. 1.27(*a*), (*b*) and (*c*)). In the absence of intervention from, for example, the float switch, the vessel in which the level is to be controlled will overflow when the exhaust valve is closed, and the boiler, isolated by valves in flow and return, explode long before self-regulation has made itself felt. Points 1 and 2 on the response curve in (*c*) show the range of values occurring during normal operation. Similar situations occur when the fan in series with an electric-or oil-fired air heater is switched off. In some cases the controller will not be capable of reacting to such emergencies; extra protective measures will then become necessary in the form of limiting instruments of one type or another.

1.5.2. Negative aspects of self-regulation

Though the presence of self-regulation is usually an asset from the point of view of control quality, this is not necessarily so. Let the position of a control valve be changed as a result of a signal from the controller. If it is opened up more, the rate of flow will increase, but not in proportion to the enlarged opening, as the self-regulating effect of series resistances will resist and limit the change. The way a change in rate of flow affects the circulating head in installations with natural circulation described above will also impede the action of a control valve. The increase in heat released per unit volume of hot water flowing through a radiator when the rate of flow is reduced is a further example of a self-regulating effect opposing the action of a controller. Many more examples could be given to show that self-regulation is not always an unmixed blessing. However, as stated, its presence is generally conducive to good control.

1.6. Control range or span; the potential values

To determine these quantities for a control system, one should close the final control element, e.g. the valve, and note the value of the controlled condition at which it comes to rest. Now open the valve wholly and note the amount by which γ_0 changes. The initial and ultimate values of γ_0 are called the potential values and are dependent upon load; the difference between the two is called the *control range* or *span*, *Sp*, as it defines the range of values within which any deviation occurring can be removed by positioning the valve accordingly. The 1970 IHVE Guide refers to this as the *controlled condition differential.*

The larger the control range, the greater the effect of changing the valve position by a given part of its stroke and the greater the chance of an over or undershoot when the controller is prevented from intervening owing to the presence of transfer lag or effective dead time. The control range is also load dependent; in Section 1.5.1, it was inferred that *Sp* increased as the rate of discharge was reduced. If, therefore *Sp* is included

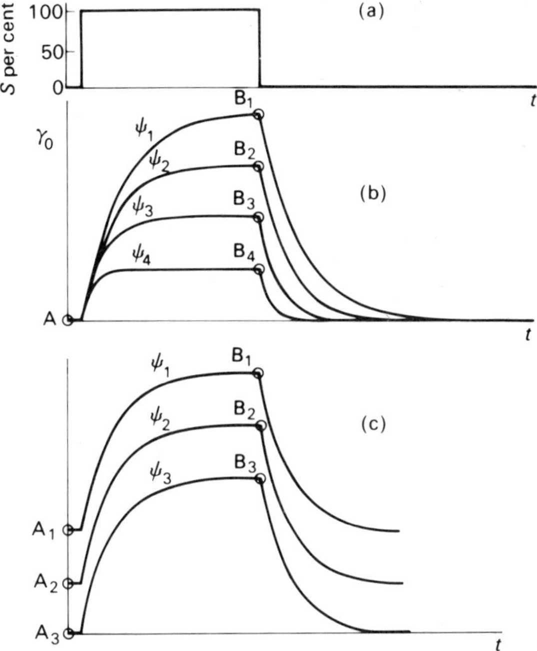

Fig. 1.28
Both control range and potential values may be affected by load, ψ.

among the control 'constants' of a process, it should be realised that, in common with the time constant and the effective dead time, its value is only constant for certain clearly defined conditions – for the rest it may vary considerably.

Figure 1.28 shows how control range and potential values may be affected by load, ψ, in the case of an air heater. At (*a*) the closed valve ($s = 0$) is seen to open up completely ($s = 100\%$) and remain open until γ_0, at B, has reached a new stable value, after which it closes once more and returns to $s = 0$. At (*b*) it is assumed that the initial temperature of the air flowing through the heater is constant, whereas its quantity varies. The initial potential value, *A*, is not affected, but both the control range and the final potential values, at *B*, are, and to a considerable degree. Once the valve is closed the variable returns to A. At (*c*) the air temperature is changed instead of its velocity: the potential values, A and B slide up and down the ordinate, the control range remaining practically unchanged. The control range also depend on factors other than load. Thus the potential water level in a vessel supplied from the mains will also depend on the mains pressure.

In many industrial processes, steps are taken to control these and similar factors, so that *Sp* may, in fact, become a control 'constant'. The larger the control range, the greater is the rate of change of the controlled variable when the controller intervenes in order to counteract a deviation. Therefore, as will be elaborated on in later chapters, a large control range is not conducive to good control quality, and it is unfortunate that, in heating and air-conditioning installations, its value cannot be chosen at will. Thus, for an air heater, the control range is given by the two potential values. The lower one may be limited to, for example, 18°C owing to the necessity of avoiding draughts and so on. The other may be limited to 40°C approximately by the danger of stratification in space. The control range in this case, 40 − 18 = 22°C must be fully utilised in order to keep the volume of circulating air down.

In large parts of the U.K., central heating installations are calculated at design temperature, i.e. a lower potential value of 0 or −1°C. The upper potential value, at 18 or 19°C, when space heating can cease, results in a control range of 18–20°C. To this extent the control problem is easy in comparison to installations in the eastern parts of the U.S.A. (design temperatures −20 to −25°C) or in northern Scandinavia (down to −35°C) where, in view of the marginally higher space temperatures desired (20 − 22°C) control ranges of 40 to 55°C are required. It will be clear that all oversizing of heating and air conditioning installations, or indeed of any installation, should be avoided where possible so as not to endanger controllability.

1.6.1. The process control characteristic, p.c.c.

In two-position control systems (see chapter 2), only the two potential values are of interest. In continuous control systems, on the other hand, it is essential to know how the controlled variable is affected by small changes in intermediate valve positions. In other words, the shape of the locus $\gamma_0 = f(s)$ is of great importance. The information given by this *process control characteristic*, p.c.c., reflects the nature of both the process and the final control element, as well as their mutual interaction.

To measure a p.c.c. is usually a difficult and time-consuming job. Operating conditions must be kept constant and the results of the

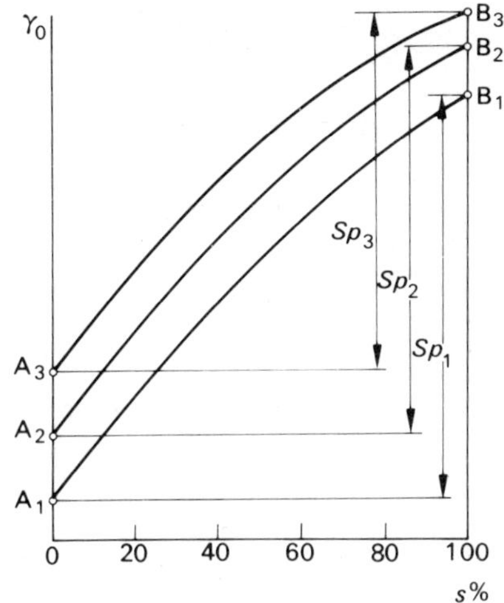

Fig. 1.29
Process control characteristics of a temperature control circuit: potential values A and B, and span Sp, may both be affected by load.

measurement corrected for any unavoidable fluctuations. Processes with long exponential and other lags take much time to reach a new stable value after a change in a valve position. Thus, rather than a line determined experimentally, the p.c.c. frequently remains nothing but a concept, albeit a concept interesting and important enough to warrant further discussion.

Though a p.c.c. may look like a response curve, as neither of them is straight and they have equal potential values and span, it should be obvious that any similarity is incidental. As shown in Fig. 1.29 a p.c.c. may be influenced by the load ψ. Figure 1.30 gives some further examples.

1.6.2. The process amplification factor, a_p

If, owing to a disturbance, γ_0 deviates from the desired value γ_i, the final control element will receive a signal characteristic of the nature and, usually, of the magnitude of the deviation γ. As a result, its action, A, will be changed in accordance with the new situation. If this change ΔA were exactly commensurate with γ, the deviation would take an infinite length of time to correct, an exponential curve being what it is. Therefore, ΔA is increased to a value several times greater. The deviation is thus corrected more quickly; as γ_i is approached ΔA is reduced.

However, in the presence of T_d, T_t or T_{de} the desired value is likely to be overshot more than once, and several subsequent readjustments of A must be accepted before the new steady state is once more attained.

Now the corrective action can be so far in excess of what was needed that a deviation with opposite sign is produced that is greater than the γ it was intended to correct. This initiates a new action which causes an even greater deviation until the final control element oscillates between its two extreme positions without ever settling down (Fig. 1.31(*e*)). The regulating system is said to *hunt*; it has become unstable. The art of the control engineer then boils down to this: how to adjust the controller so that the deviation is corrected as quickly as possible without running the risk of instability.

In this context *the static process amplification factor* a_p is a useful concept. This can be defined as the change in the controlled condition per unit change in the action of the final control element

$$a_p = \frac{\Delta\gamma_0}{\Delta A} = \frac{\gamma}{\Delta A},$$

or, in the case of a control valve,

$$a_p = \frac{\gamma}{\Delta s}$$

The qualification 'static' is added in view of the fact that when the movement of the final control element is cyclic, a_p proves to be dependent on frequency and, in non-linear processes, on amplitude also.

In pneumatic control circuits $\Delta\gamma$, ΔA and Δs can all be expressed in units of pressure – traditionally p.s.i. or bars – in which case a_p has no dimension. In electrical regulating systems it might be possible to express these factors in units of voltage or current, but as these magnitudes are usually unknown and, in any case, will differ from controller to controller, this is seldom convenient. It is therefore preferable to measure ΔA, say, in precentage valve travel and γ in units of the controlled variable. Thus a_p can have the dimension degrees Celsius per percent valve stroke (s) in temperature control, kg/m^2 per percent in pressure control, meters per percent s in level control systems, etc. The importance of the p.c.c. is now apparent, as it shows a_p as the rate of change for any value of s

$$a_p = \frac{\Delta\gamma}{\Delta s} = \tan\alpha \text{ (see Fig. 1.30)}$$

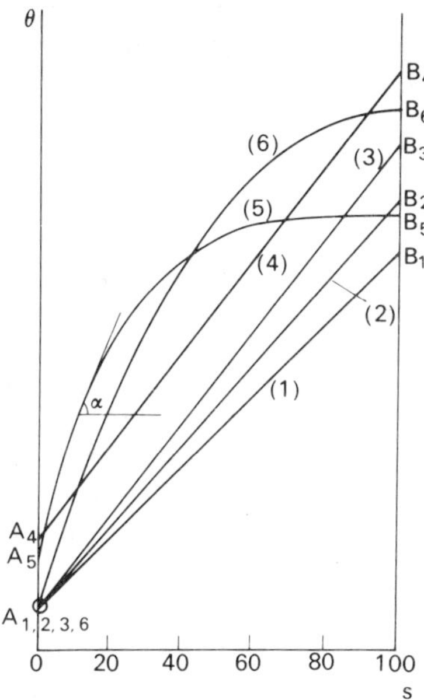

Fig. 1.30
The p.c.c. of an air heater battery at various loads and under different conditions. (1),(2),(3),(4), ideal shape due to optimum choice and sizing of all components; (5) distortion due to a badly oversized heater or valve; (6), shape as frequently encountered θ, temperature; s, valve stroke or position.

It will be clear that A will have to be adjusted to the steepest part of the curve if instability is to be avoided. This means, however, that at valve positions where a_p is less, the control action will be less vigorous than may be desirable. The best results, therefore, will be obtained when the rate of change is constant, in other words when the p.c.c. is a straight line.

1.6.3. Factors influencing the shape of p.c.c.

These are:

(*a*) The nature of the disturbance in relation to the process controlled. Temperature control systems, being essentially linear, offer the least problems in this respect. However, if in the air heater example, a given deviation is not caused by a change in fresh air temperature, but by a change in air velocity, the heat supply must vary with v^2 and with γ^2

rather than with ν and γ; the process is by no means linear. Lowering the flow temperature will reduce the slope of the p.c.c. as well as the higher potential value B.

As long as the air temperature is constant, the amount of water to be evaporated to change the relative humidity ϕ by a given percentage is also a linear function of $\Delta\phi$. This by no means applies when $\Delta\phi$ is due to a change in temperature.

(*b*) The sizing of the final control element; this subject is too vast to be dealt with here in any detail. Suffice it to say that the main factors are the static characteristic of the final control element, the mechanism by which the signal is converted into a valve movement, the medium controlled, i.e. hot water, steam, gas, etc. the, pressure distribution in that circuit, the capacity of the boiler, air heater, compressor or whatever other appliance is used in combination with the final control element to translate its action into terms of the controlled condition, the nature of the latter and the disturbances to be expected.

These important subjects will be dealt with in a companion volume[13] to this book and will not be pursued here. However, it is interesting to note that, whereas the nature of the process and the disturbances likely to occur are given, that of the valve action can, within certain limits, be manipulated to straighten the process control characteristic. In this connection the choice and sizing of the final control element are of prime importance.

(*c*) Finally, it will be clear that a_p increases in direct proportion to the control range.

1.7. Control quality

It would, at this point, seem desirable to emphasize a fact as obvious as it is frequently overlooked. The *raison d'être* of any control system is the virtual certainty that deviations will occur which, in the absence of a controller, might become too large. Efficient control action will limit their magnitude or, ultimately, even eliminate them altogether, but it can not prevent their occurrence in the first place. If no controller, however well chosen and whatever its cost, can ensure that the controlled variable will never wander from the desired value, success or failure of any control scheme must be judged by the nature of the deviations it permits.

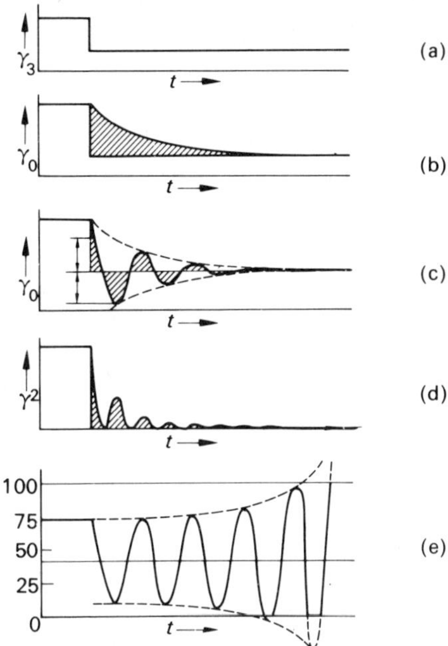

Fig. 1.31
Control quality can be assessed by various yardsticks.

So far, the term *control quality* has been used without definition. This may be excused on the grounds that no single definition can cover all aspects that, in view of the nature of the process, can be of actual or potential interest. When the band of values within which the process can operate with safety is narrow, oscillations are highly undesirable. Thus in direct expansion coils for air conditioning the upper temperature limit of the chiller surface being given by the air temperature at zero load, and the lower one limited by the danger of a freeze-up the control range is usually small. The controller will then be adjusted to give an aperiodic response (*b*) to a step change (*a*) (Fig. 1.31).Generally, a limited overshoot is permissible, provided the amplitude diminishes rapidly. In this case the control area, $\int\gamma dt$, shown shaded at (*b*) and (*c*), is a measure of the control quality. If a certain degree of oscillation is not objectionable, provided the amplitude does not exceed certain limits, the control quality can be judged on the basis of the integral $\int\gamma^2 dt$ as shown at (*d*). Finally, the time required for all oscillations to die down may be taken as a yardstick. Instability (*e*) is never permissible.

Once the definition of control quality has been agreed upon, certain types of controller are ruled out, as a study of subsequent chapters will reveal. Thus the continous oscillation of a two-step controller may be unsuitable in all cases but (*d*). The *offset*, inherent in proportional control systems, or the inability of integral controllers to take action as long as the deviation lies within the dead zone, may or may not be acceptable.

1.7.1. Controllability

In general, the above parameters are far too abstruse for practical use, and a rough and ready measure for the way a process is likely to react to automatic control is preferable. Unfortunately, it is seldom possible to determine the control characteristics, τ, T_{de} and *Sp* in the design stage. When the installation is put into operation, neither the time nor the effort may be available for a full-scale investigation of the control loop, assuming that the ultimate user is willing to bear its costs. Yet the simple method of obtaining and examining critically the response of the process to a step change can, in such cases, prove quite illuminating.

To this end γ_i is adjusted to the condition prevailing, and the process is allowed to settle down. The controller is taken out of operation, e.g. by disconnecting the valve, the position of which is then changed by hand. The way in which the controlled variable responds is then recorded and curves similar to those of Fig. 1.20 obtained. On applying a tangent to the osculating point as carefully as possible, τ_N and T_{de} can be approximated. Their ratio τ_N/T_{de}, or T_{de}/τ_N now provides a rough guide to the controllability of the process.[11]

If $\tau_N/T_{de} > 10$, the process can be controlled by simple means; at $\tau_N/T_{de} > 6$, great care must be taken if useful results are to be obtained, and if $\tau_N/T_{de} < 3$, only the most sophisticated control methods are capable of mastering the problem. Frequently one has to take recourse to open-loop methods, either as replacement of the closed-loop controller or as an aid to reduce the control range, e.g. as described in the introductory paragraphs to this chapter see Section 1.1).

In the heating and air-conditioning industry, this takes the form of a weather stat which measures one or more of the main disturbances, e.g. the outside temperature θ_0 alone, or in combination with solar radiation, wind velocity and wind temperature. The signal thus obtained is then used to modulate the flow temperature in such a way as to compensate these disturbing influences. Without interposition of further controls, all other disturbances, such as the heat generated in the controlled space by illumination, occupancy or machines, are free to affect the space temperature.

If control quality has to be improved further, a simple closed-loop controller installed to eradicate or reduce the influence of such secondary disturbances will usually prove adequate. Inasmuch as their control range is reduced, these secondary controllers operate in favourable conditions.

Apart from this, the relevance of Section 1.3.6 will now be apparent. It was shown that in control loops of higher order, the *RC* elements are of different magnitude, the smaller ones mainly influence the distortion at the foot of the response curve, whereas τ_N is effectively increased by elements with a larger time constant. Thus, it will be clear that the ratio τ_N/T_{de} may even be improved by interposition of a large *RC* element, whereas the smaller ones invariably reduce that ratio. This means that when a control loop proves difficult to control on account of its tendency to oscillate, and the circuit is examined with a view to improving control quality, attention should be mainly concentrated on elements with a small τ, i.e. exactly those that one would be most inclined to overlook.

1.7.2. Limit control

It is frequently desirable, for reasons of safety or confort, to prevent the controlled variable either from exceeding or dropping below a certain value. It may be that the danger is such that it is preferable to bring the process to an end rather than run further risks. This may be the case in frost protection where, when the low limit is transcended, the fan is switched off, the heating valve opened and a signal given. In other cases it may be preferable to allow the process to continue, trusting that the limiting device will prevent damage. Whether, in such cases, the limiting value is to be adhered to rigidly, or whether a limit ‘in depth’ is considered preferable, a discontinuity is introduced into the control system at such points which frequently, and always, in the case of a rigid limit, causes an otherwise stable control loop to oscillate to a certain extent. Provided the amplitude of the oscillation is within reason, this may be permissible, and even desirable as thus the control, temporarily transferred from the normal sensing element to the limiting device, is handed back periodically so as to prevent the limiting device from taking over control permanently. In this way, the latter can no longer maintain γ_0 at the limit once the danger of it being exceeded has passed.

This danger exists to a lesser degree in the case of a flexible, or ‘elastic’, limit, as here the detecting element never loses control completely, but is only compelled to share it to a greater or lesser degree with the limiting device. The greater the latter's 'authority' or influence, the greater the chance of instability and the poorer the control quality, in whatever sense it is used, is likely to be.

As the application of a limiting device necessarily implies that the controller can no longer maintain the controlled variable at γ_i and as in the case of a flexible limit, the same applies as far as the desired value, γ_L of the limiting device is concerned, the latter's authority, a_L can be defined as the ratio between the two departures:

$$a_L = \frac{\gamma_0 - \gamma_i}{\gamma_0 - \gamma_L} \times 100\% = \frac{\gamma}{\gamma_{LL}} \times 100\%,$$

in which γ_{LL} is the degree in which the limit is transcended and in which relationship both differences are inserted as positive values. The authority of a rigid limiting device is infinite. How to choose between rigid, and flexible limitation will be discussed at a later stage (see Section 4.6.3).

1.7.3. Average control

If the value of the controlled variable depends on the location of the detector, it may be necessary to apply corrective action in proportion to the average deviation measured in several different spots. The fact that this may well be the best solution under the circumstances should not blind one to its limitations.

If the detector only gives information on the sign of the deviation, ignoring its magnitude, it is pure numbers that count, and the average value of γ_0 will rarely conform to γ_i. Even when, as in modulating control, both magnitude and sign of the deviation are reflected in the signal, it may well be that a deviation is introduced in one part of the process where, in the absence of averaging control, γ would have equalled γ_i. In general, averaging should only be applied when the controlled variable does not differ much from spot to spot. In view of the overwhelming influence of occupancy, it would be inadvisable to make the flow of heat to a number of classrooms dependent upon their average requirements. On the other hand, in a theatre where the frequently cold and draughty stage may cause significant temperature differences between front and back of the hall, averaging may be the best solution, when other methods, e.g. causing a flow of air to, rather than from, the stage by suitably locating the exhaust air duct, or by installing some sort of an air curtain between stage and hall, are not feasible or are too expensive.

1.8. The controlling factors summarised

Summarising, it has been shown in this chapter that the main factors determining the response of a process to automatic control are: its time constant, τ_N, usually the longer the better, its effective dead time, T_{de}, which, in common with the control range, *Sp*, affects control quality more and more unfavourably, the greater it gets.

From this it would seem obvious, that, apart from their individual magnitudes, the relationship between these three control constants will go far to determine the controllability of a given process; for the ratio τ_N/T_{de} this was confirmed in Section 1.7.1. In fact, a number of diagrams have been published both by Junker[6,7] and Kromwijk,[10] in which the influence of such parameters on the results obtained with controllers of a given type has been summarised. Such graphs are valuable inasmuch as they deepen one's insight and provide a guide to action in difficult cases. For this reason a selection of them will be reproduced and discussed in subsequent chapters.

However useful and illuminating such graphical representation of highly complex relationships prove to be, one should not set one's expectations higher than is justified. Several grounds for caution were set out earlier, one of the most important being that the control 'constants' are by no means constant and may vary very considerably with load and with several other factors dependent upon the process and mode of control. The effective control range under any given load conditions especially will be influenced by the degree to which the process control characteristic departs from a straight line.

A difficulty of a different kind is the virtual impossibility of drawing a tangent with any degree of accuracy at the osculation point of an experimentally obtained response curve. This was difficult enough in the construction of Fig. 1.18, but it is far more so in the field owing to the spread of the measured points due to random influences, as in Fig. 1.20. Moreover a slight variation in slope of the tangent has an appreciable effect on the apparent magnitudes of τ_N and T_{de} and an even greater influence on their ratio. Apart from this, the difference between transfer lag and distance-velocity lag is ignored and, as Fig. 1.18 also showed, the τ_N found gave only a very rough approximation of the actual shape of the curve. The combined effect of these factors may cause significant departures from the response that the diagrams may lead one to expect. They should therefore never be used as an accurate means of forecasting results.

1.9. Some practical considerations

1.9.1. Energy supply and controller characteristics

For its operation a controller needs energy. This may be derived from the process it is controlling, in which case the controller is called self-acting. When the energy is derived from fluids like oil or air under pressure we speak of hydraulic or pneumatic controllers respectively. Finally, the controller may be energised electrically. This group is subdivided into electro-mechanical and electronic controllers. The latter term is used to denote the fact that the signal emitted by the sensing element needs amplification by electronic means before it can be used to activate the final control element.

The type of energy used is the largest single influence on the construction and on the secondary characteristics of a controller. However different in appearance the various types may be, their primary control characteristics conform to the same laws. These laws linked to a coherent if simplified theory, will be set out in the following chapters. Constructional details will only be brought up on the rare occasions when required for the understanding of the subject; they could easily fill a volume twice this size. Each type of controller has its own particular field of application and it would seem useful to devote a page or two to such features on which they differ, before discussing what they have in common.

1.9.2. Self-acting controllers

As mentioned these self-contained controllers are energised by the process to which they are applied. They are frequently cheaper than any other kind, easily mounted and have few, if any, external connections. Automatic air valves are used to de-aerate steam vessels such as boilers, radiators or non-storage calorifiers when these are taken into operation, or to aerate them to prevent the pressure from falling below atmospheric. If they discriminate between air and steam by the difference in temperature or density, these are also the agents that open or close the valve through which the air passes. Valves of this type need no external connections at all. The same applies to pressure relief valves and in many cases, to reducing valves. Self-acting temperature control valves frequently are connected to the sensing element by a capillary tube. In such cases application is limited by the capillary length, which for several reasons, has to be kept as short as possible.

The great majority of self-acting controllers are of the proportional type (see Chapter 4). Though a local adjustment of the value at which the variable is to be kept constant is usually essential, they cannot be reset from a control panel or other central point. The proportional band is restricted by the design and can rarely be adjusted to suit the characteristics of the process.

The force available to change the valve position within its stroke is limited so that their application is inadvisable in all cases where they have to move against an appreciable differential pressure, or where the friction to be overcome is likely to increase with life due to corrosion, sludge or other contaminants – such as is frequently the case in chemical processes. Also, the valve stroke is usually restricted; both features combined make the use of double-seated valves very attractive. However, owing to their simplicity and low price it is generally worth checking their suitability for any automatic control application with which one may be concerned.

1.9.3. Hydraulic controllers

These are generally operated by pressurised oil. They were developed for use with heavy machinery such as steam engines, internal combustion motors, large lathes, and other metal processing equipment, where oil was used in any case for lubrication and cooling purposes.

In their simplest form hydraulic controllers are of the integral mode (see Chapter 3). Their main feature is the immense force available for moving the final control element, limited as it is, only by the mechanical strength of the component parts. The oil circulates in a closed circuit as, once it has done its work, it must be returned to the pump. Thus it is at a disadvantage compared to pneumatic controllers where the compressed air is exhausted to the atmosphere after having fulfilled its function.

However, the wholly enclosed circuit makes hydraulic control systems eminently suitable for applications in dusty atmospheres such as in stone processing factories, mines, cement works, etc, where pneumatic controllers would soon get clogged. If there are leaks in the tubing and past the connections, this is not necessarily disastrous, as long as the reliability of the control system is not endangered. In many cases the inflammability of the oil is a serious handicap; the search for alternative fluids has so far not proved particularly successful.

1.9.4. Electro-mechanical controllers

Electrical energy is universally available wherever automatic controls are likely to be used, and skilled electricians capable of installing them are to be found in practically every locality. Thus their application is particularly attractive in small installations such as the temperature control in houses, etc. As, at the same time, physical quantities are readily converted to electrical signals which can be conveyed without noticeable delay over almost any distance; use is equally attractive for large installations controlled from a central point. Relays provide a simple means of linking two or more separate control circuits, and disturbances can give rise to visible or audible signals generated by simple equipment.

1.9.5. Electronic controllers

These (see Chapter 5) share many features of electro-mechanical controllers (see above). They are, or can be, extremely sensitive and free from hysteresis. Thus, wherever a high degree of accuracy is required, electronic controllers are likely to be favoured. Any mode of control can be realised by simple electronic circuits, and stabilising factors can be incorporated by similar means. The sensing elements are energised by low voltages and few, if any, switching operations have to be carried out in such circuits. Thus, wherever there is a limited danger of explosion, they have advantages over electro-mechanical types. In processes however, where the danger of explosions is acute, their use is inadvisable.

1.9.6. Pneumatic controllers

This type (see Chapter 6) has a great advantage over both electrical types inasmuch as electric motors are much more complicated and expensive than pneumatic actuators. The price difference is considerable and the savings in extensive control schemes in which valves etc are used in numbers will easily offset the price of the compressor and associated equipment. Pneumatic actuators also have some very real technical advantages; as spring return is an inherent feature they are safer, and the force available to position the control valve or damper can be increased far beyond that available from comparably priced electric motors. Thus, the actuator is not, or need not be, the factor limiting the differential pressure over a valve or the air velocity through a damper. Their initial reaction to a disturbance can be very fast. Yet they slow down as the new position is approached, thus reducing overshoot and increasing stability.

The range of air pressures is practically standardised internationally to 0·2 — 1·1 bar, or 3 — 15 p.s.i., a great advantage when replacements have to be found in a hurry. Many types of sensing element carry their own amplifier; thus pneumatic control panels can be smaller and cheaper than their electrical counterparts. Special systems allow pneumatic signals generated for control purposes to be used for measurement or indication on simple calibrated manometers without detriment to their normal control function. On the other hand, the latest tendency to control very large installations by a computer is more easily realised electronically than by pneumatics.

1.9.7. Electro-pneumatic controllers

The rapid response and accuracy of electronic sensing elements and the low price of pneumatic actuators are combined in this system. Unfortunately, thus far, the price tends to be high for universal application.

1.9.8. Controllers for heating, ventilating and air conditioning installations

For small oil or gas fired central heating installations, electro-mechanical controllers are the obvious choice on account of their simplicity and price. In recent years the price of electronic controllers has dropped to such an extent that their use is becoming widespread in somewhat more sophisticated installations.

In large heating plants all types of controller with a possible exception of hydraulic types are found, practically in equal proportions. However, the majority of large air-conditioning installations are controlled pneumatically, even though a sprinkling of other types is invariably present, frequently for special purposes. Thus, to give a few examples, a self-acting controller may be chosen for a storage calorifier; the frost protection thermostat will usually be of the electro-mechanical type as it has to switch off the power supply to the ventilator when a dangerous situation arises; and the influence of sun and wind is conveniently detected electronically.

2. Two-positional (including on-off)and multi-positional control

2.1. General

According to B.S. 1523, a two-positional (including on-off) controller is a controller in which the output signal changes from one predetermined value to another when the deviation changes sign. Thus, the final control element can only come to rest in one of two positions, usually, but not necessarily, the extreme positions; both on-off and high-low control systems are two-positional controllers (Fig. 2.1 (*a*) and (*b*)). In the latter case the valve may be bypassed, or may remain slightly open in the 'low' position, in which case it will not close further unless overruled by the signal from another detector. This could occur when the load drops beyond that corresponding to the low positon, as a result of which the controlled variable rises to a potentially dangerous value. In this case the system as a whole can be described as a three- or multi-positional control system. In general, the valve position, or rather that of the controlling contact, indicates the sense in which γ_0 departs from γ_i. However, the values of γ_0 at which the valve is made to assume its alternative position overlap somewhat, and therefore the switch-over point on a falling γ_0 is lower than that at a rising value of γ_0. Within the narrow range of values set by the overlap, γ_d, or differential as it is usually called, and apart from the question as to whether γ_i has been arranged to correspond to one of the switch-over points or to an intermediate value, the relation between γ_0 and γ_i will depend upon previous history. Thus the position of the final control element can never indicate the magnitude of the deviation. Two-positional controllers can be actuated in any way desired, but it is not a mode for which pneumatic or hydraulic controllers are naturally suited in the way electro-mechanical controllers are. The most popular two-position device and, at the same time the prevalent electro-mechanical controller, is the simple electrical switch with snap-action operated by, or at least in sympathy with, the controlled condition. It is with devices of this kind that two-position control will be illustrated in what follows.

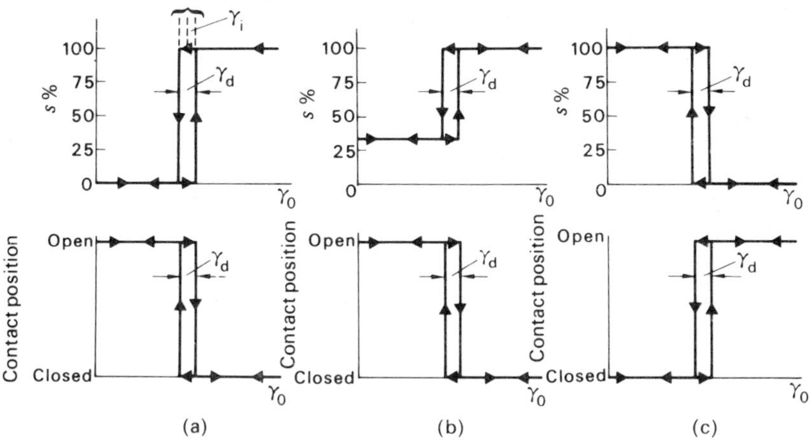

Fig. 2.1
Electrical two-positional control: Contact- and valve positon, s, as a function of the controlled variable. The desired value, γ_i, need not necessarily be central to the differential, γ_d. (*a*) on-off (*b*) high-low control, both for heating; (*c*) on-offcontrol for cooling processes.

S.p.s.t. contacts may switch an electric heater on or off, open or close a solenoid valve, or energise an oil or gas burner relay. Alternatively, the switch may be fitted with an s.p.d.t. contact with which to operate a motorised valve or damper. Special motors made for the purpose run either 180°, and a further 180° as the contact changes over, or backwards and forwards over the same, arbitrary, angle.

2.1.1. The type of control action

Providing it is properly applied, two-positional control can given satisfaction when judged by the following criteria:

(*a*) The deviation, γ, from γ_i should never exceed a given maximum value.

(*b*) The relationship of the mean value of γ to the two switching points should vary as little as possible.

(*c*) From a point of view of wear, oscillating frequency should be low. As, on the other hand, control quality is usually improved by increasing the frequency, cycling time is sometimes reduced artificially, wear permitting.

2.1.2. Peculiarities of two-positional control action; the sustained deviation

Figure 2.2 shows an idealised first-order process with and without dead time in which time constant τ and effective dead time τ_d are supposed not to be affected by the phase of the control action ('charge' or 'discharge'). γ_d has been enlarged in relation to *Sp* in order to show its influence more clearly. At (*a*) the load is assumed to be 50 per cent of design load, and the way in which γ_0 would vary in the absence of a controller when either of the potential values is chosen as point of departure, is shown by curves (i) and (ii). Starting from the lower potential value (1) γ_0 will increase until, at point 2, the upper limit of the overlap γ_d is reached. In the absence of any dead time, γ will immediately start to drop in accordance with the corresponding part of curve (ii). At point 3, the line reverses once more, etc; the result is a sawtooth line of which the amplitude equals $\gamma_d/2$ and the mean value, $\gamma_m = \gamma_0$. The shaded area on either side of the γ_i line are congruent.

As the load increases (*b*), the potential values will drop along the ordinate with respect to γ_i, and the figure becomes asymmetrical. The shaded areas are no longer equal and the mean value γ_m rises with respect to γ_i by a small amount γ, a deviation which, for a given load condition, is sustained. The reverse happens as the load decreases (*c*); γ also reverses its sign. As the figure becomes more asymmetrical so, also, the cycling time *T* will increase.

Such effects are enhanced by the introduction of a time lag, shown for the sake of clarity as a pure velocity-distance lag T_d. More important, the amplitude of the oscillation *A* is increased out of all proportion: γ_0 is no longer confined to the differential but swings well beyond its limits, and the more the load departs from the 50 per cent mark, the greater both the deviations and the oscillation time become.

Figure 2.3, in which τ has been increased with respect to *Sp* and T_d, shows a considerable improvement when compared with Fig. 2.2. The increase in cycling time is also striking. Thus we can expect the results obtained with a two-step controller to improve with decreasing ratios T_d/τ and Sp/τ.

Some important conclusions can be drawn from these figures. The greater the rate of change $d\gamma_0/dt$, the larger the deviation is likely to be for a given T_d. As the slope increases, with shorter τ and greater Sp_0 values, it can be said in general that the results obtained with two-positional control will be better the longer τ_N, the shorter T_{de} and the smaller *Sp*. The fact that, in Germany, the *Sp* for heating installations is

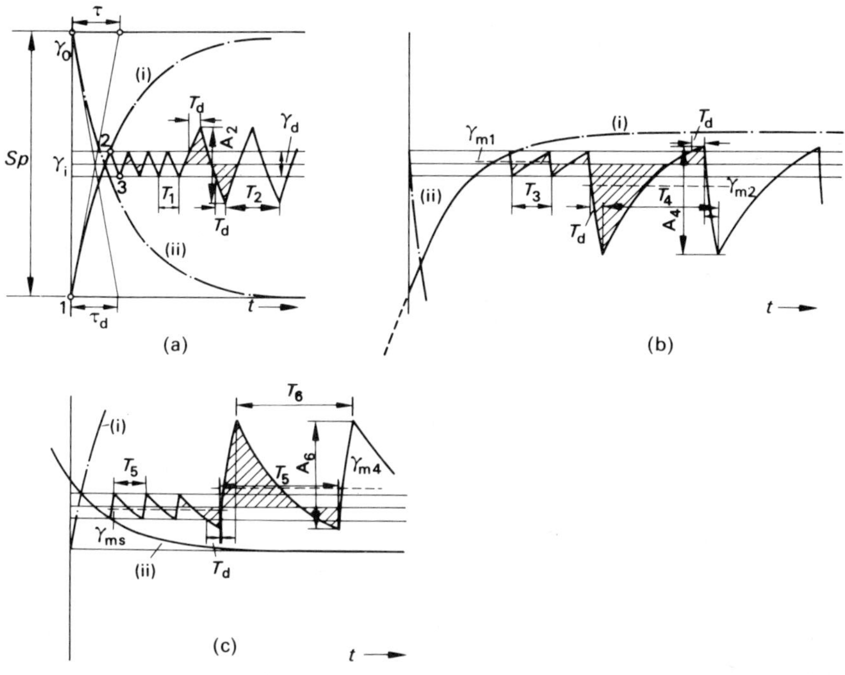

Fig. 2.2
Idealised response curves. (*a*) for mean load; (*b*) for high load, and (*c*) for low load, with and without true dead time. The values of τ are assumed to be equal for both phases. γ_d, overlap; T_1, T_2 etc. cycling times; γ_{max} amplitude; T_d true dead time; *Sp* control range.

nearly twice the English value is one reason why two-step space temperature control is far less popular over there than it is here. Another striking point is that the sustained deviation mentioned earlier is completely swamped by the effect of T_d which reverses its sign and increases the magnitude of the sustained deviation more or less in proportion to the amount by which the load deviates from the average value. A similar tendency will be apparent when, as is frequently the case, τ is shorter and *Sp* is larger in the charge phase, than in the discharge phase.

Figure 2.4, referring to a process of the first order with true distance-velocity lag gives a more specific picture of how these 'constants' are reflected in the results.[6] All parameters are dimensionless. Along the abscissa the scale is marked up for the ratio of two times, T_{de}/τ rather than the reciprocal which has been used until now. There are two inde-

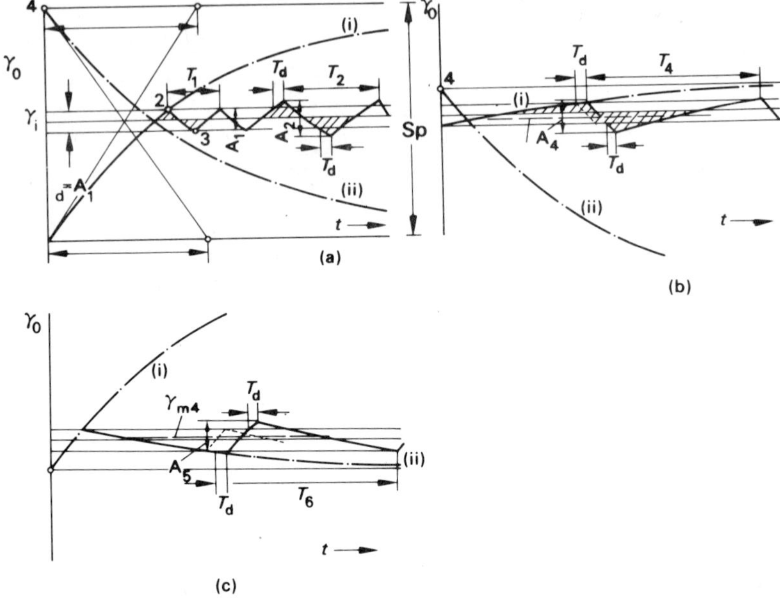

Fig. 2.3
As Fig. 2.2 with larger values for τ.

pendent scales along the ordinate: the left-hand scale indicates the ratio between the peak to peak value of the oscillations, A on the one hand, and the control range Sp, i.e. the distance between the potential values expressed in terms of the controlled condition on the other. Various curves are given with the ratio of overlap to span γ_d/Sp as parameter.

On the right-hand side a subsidiary scale is drawn for the product of frequency and time constant, $f\tau$, once more a dimensionless magnitude. The diagram refers to conditions as set out in Fig. 2.2 (*a*), i.e. for the case when the oscillation is symmetrical to a mean value, and that τ and T_{de} are the same for the charge and discharge phases.

Take as an example a storage calorifier having the following parameters: $\tau = 30$ min, $T_{de} = 0{\cdot}6$ min, $Sp_0 = 60 - 15 = 45°$C; the thermostat used may have a differential of $4{\cdot}5°$C. Thus, $T_{de}/\tau = 0{\cdot}02$, and $\gamma_d/Sp = 0{\cdot}1$. For the ordinate of the point of intersection we find $\gamma_{max}/Sp = 0{\cdot}11$ approximately on the left-hand and $f\tau$ a little more than 2 on the right-hand scale. Thus γ_{max} proves to be 45×0.11 or $4{\cdot}55°$C approximately, and the frequency 4 (per hour). Both values are quite acceptable but, as mentioned, refer to 50% load only.

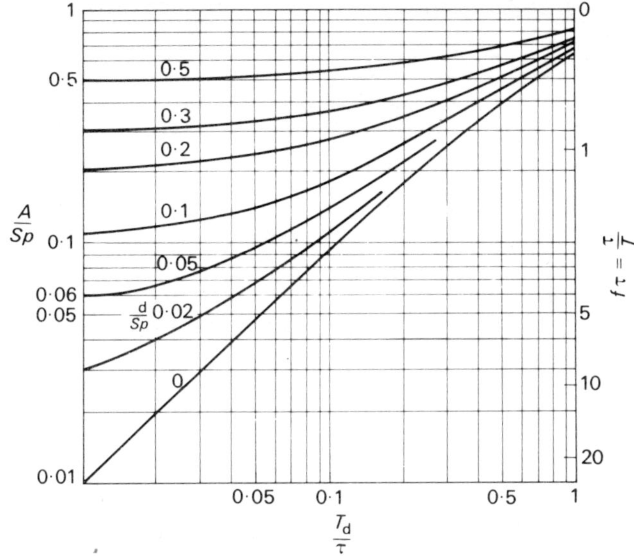

Fig. 2.4
The relationship of amplitude $A/2$, overlap γ_d, control range Sp, (true) dead time T_d, time constant τ and cycling frequency f for single-stage processes with dead time controlled by a two-positional regulating system at mean load according to Junker.[6]

As a contrast, take the case of a roomstat controlling an old-fashioned boiler. The constants now change to $\tau = 1$ hour, $T_{de} = 6$ min, $\gamma_d = 1°C$, $Sp = 20 - 0 = 20°C$ and the ratios T_{de}/τ and γ_d/Sp are 0·1 and 0·05 respectively. We then find $\gamma_{max}/Sp = 0{\cdot}14$, or $A = 0{\cdot}14 \times 20 = 2{\cdot}8°C$, and $f\tau = 1{\cdot}7$ or $T = 35$ min. As a room temperature should normally not vary by more than $\pm\frac{1}{2}$ to $\frac{3}{4}°C$, an amplitude of $\pm 1\frac{1}{2}°C$ is too large, especially when secondary effects such as fluctuations in radiant heat from the radiator and the '*cold 70*' effect in the room are also taken into consideration. This difficulty is effectively overcome by increasing f artificially by means of heat acceleration (see Section 2.3.1.). If f is multiplied no more than three times, $f\tau$ becomes approximately 5 and γ_{max}/Sp becomes 0·05. Thus we find $A = 20 \times 0{\cdot}05 = 1°C$ and the amplitude is reduced to $\pm\frac{1}{2}°C$. Usually the frequency will be increased considerably more than three times by heat acceleration, and the amplitude will decrease roughly in proportion.

2.3. Typical applications of two-positional control

Without going into detail, this simplest of control modes is most suitable for applications in which some or all of the following requirements are met: the control range must be small, the lags short and the time constant long; also, fluctuations in the controlled condition must not be objectionable. Thus, burners in boilers are commonly controlled on-off or high-low, as well as boiler-feed processes and the temperature control in non-storage calorifiers.

As explained in Section 1.8, the values obtained from Fig. 2.4 are indicative of trends only. Why the magnitudes will not correspond exactly to values obtained in practice will be apparent from a closer examination of the 'constants' as obtained, for example, in heating applications.

(*a*) τ for the charge and discharge phases are seldom equal (Section 1.3.5). The same applies to T_{de}: as the average temperature of a radiator increases, so the air circulation in space is increased and T_{de} decreases.

(*b*) The figure refers to 50 per cent load only.

(*c*) On account of the usual oversizing of a heating installation, the effective control range is invariably greater in the charge than in the discharge phase.

(*d*) The values for τ and T_{de} interpreted from a response curve found experimentally are but approximations of the values upon which the diagram is based (Section 1.8).

As a result the symmetry assumed in Figs. 2.2 and 2.3(*a*) is lost, and the point at which the direction of the curve is reversed is rounded off (Fig. 2.5).

When discussing (Fig. 2.4), the conclusion was reached that special measures are required to make two-position regulators suitable for space-temperature control, and this is confirmed by these examples. Such measures are:

(*a*) Reduction of the T_{de} by choosing boilers and radiators with as small a volume of water as possible; by increasing the water velocity and thus using smaller pipes, both distance-velocity and transfer lags are reduced, as is the case by using diverter valves for zones situated far from the boiler; locating the thermostat in a position where changes in room temperature are readily detected; and choosing a sensing element with as small a τ and T_{de} as possible.

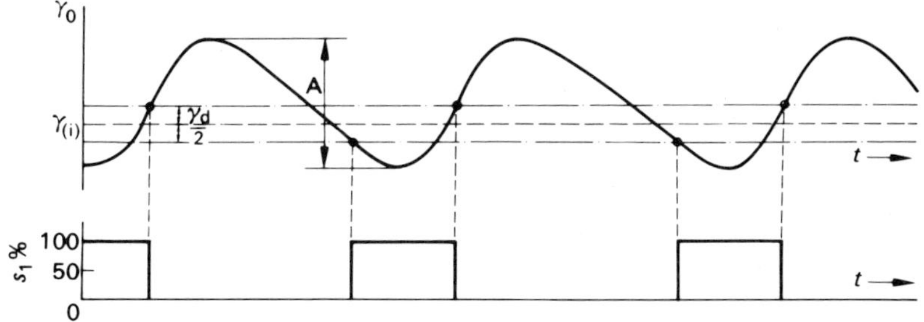

Fig. 2.5
The effect of transfer lag and unequal control range and time constants at 50 per cent load. *s* represents valve position, other symbols as in Fig. 2.2.

(*b*) Reduction of the control range by sizing burner, boiler and radiators as small as commensurate with meeting the maximum load, or, alternatively, by applying a primary flow-temperature control by means of a 'weather controller' or compensator (see Chapter 9).

Reduction of the differential below a value reconcilable with its function as a switch has little point in comparison with the other factors mentioned. Heat acceleration, on the other hand, is an extremely valuable method of improving results.

2.3.1. Heat acceleration

By means of heat acceleration, the temperature of the temperature-sensitive element is raised beyond ambient every time, and as long as, the thermostant calls for heat. The moment the switch-on contact closes, the valve opens, the burner, or whatever other final control element is used, comes into operation. As the radiator fills with hot water, convection currents develop in the air, heat spreads slowly throughout the room. The considerably attenuated heat front eventually arrives at the sensing element. The temperature of cover and base, as well as that of the sensitive element, rises slowly. When at last the switch-off temperature is reached and the heat supply is cut off, the radiator is full of hot water. This heat must be dissipated, causing a further rise in space temperature before the action of the thermostat is effective. As a result, the space temperature will overshoot its mark although the thermostat took the action designed to prevent this a time equal to T_d earlier. It is due to this dead time that the period during which the thermostat calls for heat is longer than needed to satisfy the demand:

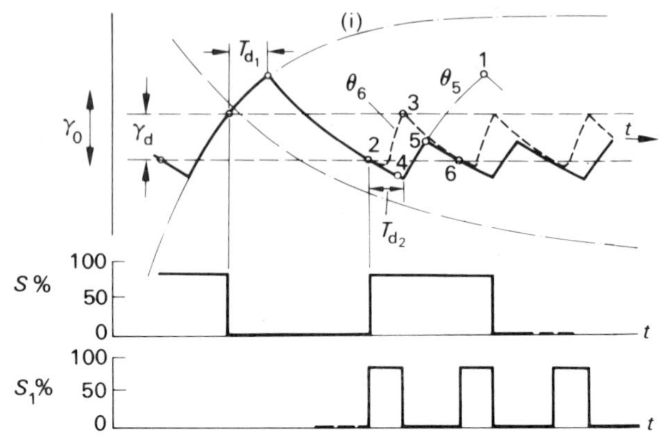

Fig. 2.6
The action of a two-step thermostat improved by heat acceleration. (*a*) The zig-zag line shows the space temperature, θ_s; the dashed line that of the sensing bimetal element, θ_b, first without, then with heat acceleration; (*b*) and (*c*) The switching frequency increases when acceleration is introduced.

the shorter τ and the greater *Sp* of the process, the greater the overshoot will be.

Though heat acceleration cannot in itself change all this, it can do much to improve control quality. In heating applications the effect is achieved by warming up the sensing element (e.g. electrically) during the demand phase, as a result of which the thermostat will switch off the heat supply earlier than if no accelerating element were present. Referring to Fig. 2.6, without heat acceleration the temperature would rise to point 1. At point 2, however, the accelerating element comes into operation causing the temperature of the sensing element to rise steeply and to stop the flow of heat at a moment, 3, when the space temperature is still represented by point 4. Due to the presence of dead of dead time it subsequently rises to point 5 before its direction is reversed. Now two effects occur simultaneously. The detector cools down slowly on account of the small temperature difference between it and its surroundings. The delay thus introduced enables the heat stored in the radiators to spread, as a result of which the temperature of the air surrounding the detector will rise to meet that of the latter. Dependent upon the load, the switch-on temperature is reached sooner or later (6), and the process repeats itself. Heat acceleration thus fulfills several duties:

(*a*) it reduces the cycling time and, with it, the amplitude of the oscillation

(*b*) it cuts out the effect on the sensing element of the process dead time, replacing it by the far shorter time taken by the heat transfer by direct contact from heating element to sensing element

(*c*) it allows the differential γ_d to be adjusted – to any value necessary – for efficient switching action without appreciably affecting control quality

(*d*) 'cold 70' effects are reduced. In early spring when the nights are still cold the thermostat will register a prolonged call for heat first thing in the morning in order to replace the heat lost during the night. Thus the radiators will be filled with hot water. The sun will shine through the windows and cause the temperature in the room to rise more rapidly than that outside. Thus it will be a long time before more heat is required. The convection currents above the cooled down radiators come to a standstill, but the inside air, chilled at the windows, will continue to drop towards the floor upon which a stationary layer of cold air will soon form. The occupants complain that it is too cold even though a thermometer placed next to the thermostat at a height of 1·5 m, may indicate 21°C. Heat acceleration chops the protracted demand for heat into a number of shorter periods; the overshoot is less, and the radiators, heated more moderately, maintain the air circulation and warm up the air cooled at the windows for a longer period. Circulation is maintained and the 'cold 70' effect is avoided or diminished.

Against these considerable advantages one unpleasant side effect must be set. Heat acceleration superimposes a proportional component upon the two-step regulation scheme. As the load increases, the charge periods will necessarily become longer, the accelerating element will be energised for longer periods at a time, and the bimetal temperature will be raised further in relation to ambient. The switch-off point being determined solely by the bimetal temperature means that, as the load increases, the flow of heat will be cut off at progressively lower ambient temperatures.

This off-set at maximum load can be reduced to a few degrees C by effective design: the *R* and the *C* of the bimetal heater assembly can both be decreased by a close contact between the two, and by using as light a bimetal as is commensurate with a reliable switching action. Since, in heating installations, sudden load changes seldom occur, this offset means at the worst an occasional re-adjustment of the desired temperature with changes in load, a low price to pay for a very real improvement in control quality obtainable with the simplest of controlling devices.

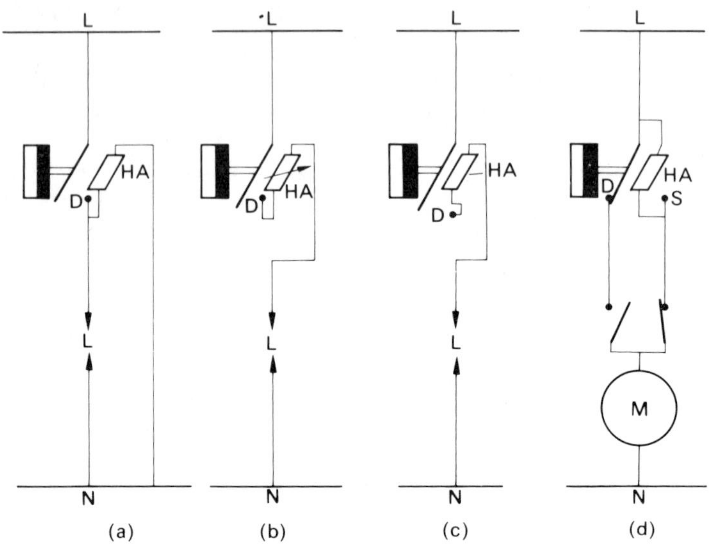

Fig. 2.7
Simple methods of providing heat acceleration (HA) (*a*) (*b*) for a load *L* of arbitrary magnitude; (*c*) (*d*) for a load of clearly defined magnitude. Contacts marked D when closed, denote a demand, or a call for, say, heat; S-contacts closing show that the demand has been satisfied.

2.3.2. Adjustable heat acceleration

It might be thought desirable to adjust the amount of acceleration to the magnitude of γ_d, Sp, U and T_d. In practice a compromise solution has proved to be quite adequate. However, adjustability may be provided, but the for a wholly different reason. Figure 2.7 shows the four simple methods by which heat acceleration is commonly achieved. When the switch contact closes at (*a*) the accelerating element is energised; the amount of heat it generates is independent of the nature of the final control element which may be a relay, a solenoid valve or a large electric heater. At (*b*),(*c*) and (*d*), the load current also flows through the accelerating element. Its resistance R can then be adjusted to the magnitude of the current I to be expected (*b*) so as to provide the correct amount of heat dissipation $Q = I^2R$, the scale being calibrated in amperes. Alternatively, it may be assumed that thermostat and relay (*c*) or motor (*d*) are of the same make and will always be used in known combinations, so that the resistance and rating of the heat accelerating element can be determined in advance and adjustment is not necessary. Actually solutions to (*b*),(*c*) and (*d*) are only feasible at low voltage. At mains voltages the very small current would call for a high resistance $R = V^2/Q$, and the resultant voltage drop I/R would be excessive. Frequently, heat acceleration is provided by natural means. If a two-position thermostat controlling the temperature of the flow of recirculated water in a swimming bath is placed immediately after the heater battery, the fluctuations can be kept to a fraction of the overlap by the frequency with which the battery is switched on and off. Again, in a hot-house, a stem-type of sensing element may have to be used in preference to a spacestat, as these are seldom watertight. If the phial or stem is suspended in the convection current above the heating coil, fluctuations in air temperature can be kept to a minimum. A similar effect is achieved by soldering the pocket of a stem thermostat to the bottom of some receptacle, e.g. a degreaser that is heated from below. For pneumatic or hydraulic controllers, such methods are the only ones available.

2.3.3. Automatic adjustment of heat acceleration

In the Scandinavian countries or Canada, where the *Sp* in heating applications can be very large, the offset due to heat acceleration may become excessive. For those cases a device called an inside-outside controller (Fig. 2.8) has been developed. It consists of an outside detector, OE, the housing of which is partly transparent so as to make it sensitive to solar radiation. It contains a double-pole switch actuated by a bimetal, one pole of which controls a heater capable of keeping the outside element at, say, 20°C, independent of weather conditions. As it gets colder the second switch contact, wired into the connection between the heat acceleration element in an indoor thermostat, IE, and neutral, will remain closed for longer periods, thus eliminating the offset to the extent that the load arises from weather conditions.

The inside-outside controller, by the way, is an example of an open-loop element used to correct, or at least moderate certain shortcomings of a closed-loop control system.

2.4. The block diagram

Figure 2.9 is a diagrammatical representation of a two-positional control loop. At (*a*) the characteristics of the process are shown, summarised by time constant, dead time and range, τ, T_d and *Sp* respectively. At (*b*)

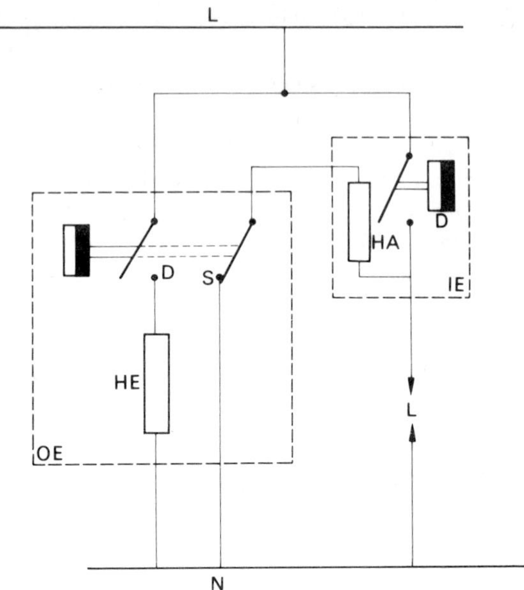

Fig. 2.8
Inside-outside controller. OE: outdoor element; IE: indoor element; HA: heat accelerator; HE: heating element; *L*, load. The closing of the D- or S-contacts respectively denote a call for, or a sufficiency of, for example heat.

the mode of operation of the detector is indicated by the two switching positions located at a distance γ_d from one another. Detail (*c*) shows that the final control element is assumed to operate at high speed, or at least so rapidly that the time taken to reach an alternative position is negligible in comparison with the oscillating time. Finally, at (*d*), the result of the control action is shown, characterised by the double amplitude, *A* and period, *T*, of the oscillation. Disturbances u_1, u_2 etc may enter the loop at various places and affect γ_0 accordingly.

2.5. The practical application of two-positional control

Though a discussion of constructional details would exceed the scope of this book, a few words on the use of two-position control will round off this chapter. The characteristics of processes particularly suited to this mode were described in the course of earlier Sections.

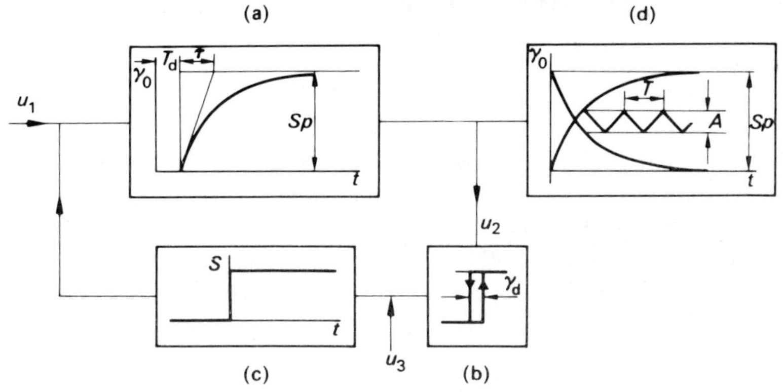

Fig. 2.9
Block diagram of a two-step control circuit. (*a*) the process controlled. (*b*) the sensing element. (*c*) the final control element. (*d*) the result of the control action, all characterised by their features.

2.5.1. The set value

For calibration purposes, a point located centrally within the overlap would seem the obvious set value. However, if, apart from the aspect of control, the emphasis is on preventing the controlled variable from exceeding a certain safe value, the set value may well be chosen to coincide with the upper switching point. On the other hand, application of heat acceleration will tend to lower the average value of γ_0. So as to split up the offset into two small deviations, one positive and one negative rather than one large minus droop, the thermostat is calibrated to a lower value, e.g. to the switch-on point. As periods of extreme cold are rare and of relatively short duration, the emphasis will be on conditions at light load, and it may, at heavy load, be desirable to adjust a roomstat to a value higher than actually required so as to compensate for the offset. The same applies when the thermostat is mounted on a cold wall by mistake. Alternatively, it may well be impossible to locate the sensing element in such a way that it can measure the variable accurately. Thus, in a cooking oven, a phial placed in the centre would be in the way and be damaged by pie dishes, etc. Along the wall, where it is out of the way, the temperature will be lower by an amount increasing with the oven temperature. By miscalibrating the thermostat in such a way that switching occurs at a phial temperature, say, 20°C below the set value, the error is reduced and accuracy sufficient for the purpose is achieved.

2.5.2. Limit control

In a two-position scheme, the action of any limiting device must necessarily be of the same kind. Thus, in order to prevent an excessive temperature, a signal to open a valve will be converted to one closing it. As a result, the temperature will drop once more, the limiting device will be satisfied and the control will be returned to the sensing element. As the intervention of the limiting device will have increased the deviation, the valve will immediately open up again. Thus a two-positional limiting device will necessarily cause a continuous opening and closing of the valve with fluctuations in the controlled variable that may well be felt to be unacceptable (Fig. 2.10).

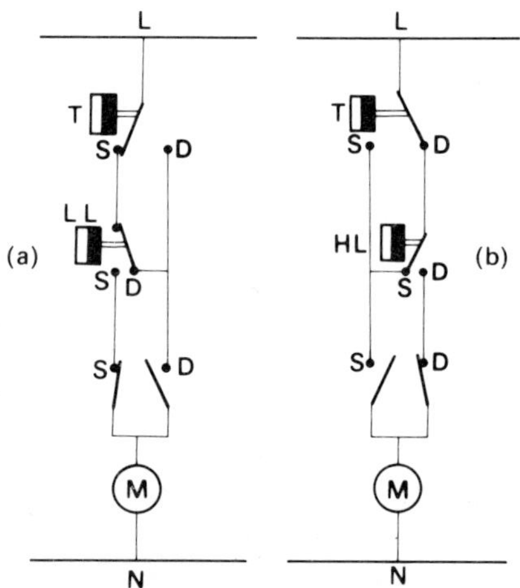

Fig. 2.10
Arrangement of limit-stats in two-step control systems; (*a*) low-limit control, (*b*) high-limit control. The position of its limit switches shows that the motor has already opened or closed in accordance with the command of the limiting device.(*See next section*)

2.5.3. Sequencing and following control

On account of limitations on the permissible differential pressure, or for other reasons, one large valve may have to be replaced by two smaller ones operated together. Also, two dampers that should move simultaneously or nearly so may be located too far apart for them to be operated by a single motor, which may or may not be powerful enough for that duty. In such cases one motor is made to follow movements of the other with a minimum of delay. Alternatively, in two-stage schemes the heating valve must close before the cooling valve may open. In sequencing valves the use of two separate actuators, optional in the other cases described, is essential. Yet the signal is likely to be derived from a single contact every time.

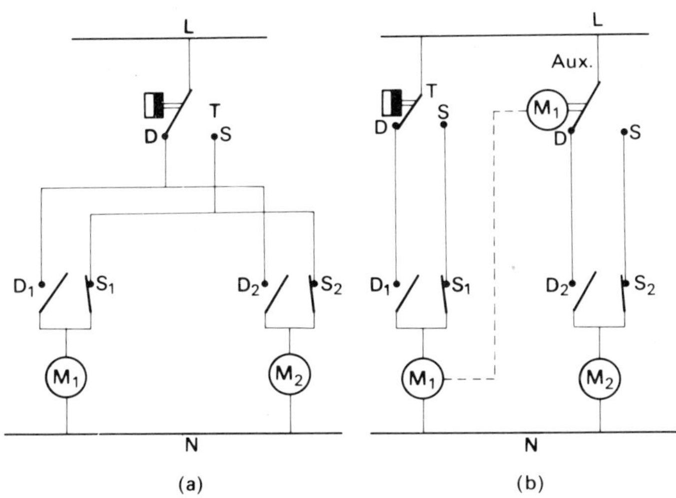

Fig. 2.11
Sequence and follower control of two motorised valves or dampers in two-step control systems; (*a*) wrong method (*b*) correct method.

Should the control scheme be arranged as in Fig. 2.11 (*a*), trouble is likely to occur due to small variations in motor speed. When the demand is satisfied, the S-contacts of the controlling instrument will be closed. Those of the motors, on the other hand, will be open and the D-contacts closed. At a changeover of the controlling contact. both motors will start up. Soon after, their S-contacts will close in readiness for emergencies. If now M_1 reaches its new position marginally earlier than M_2 and D opens, it will continue to run on as a circuit is maintained from D, though D_2, S_2 and S_1 to M_1. A moment later when D_2 breaks M_2 will also continue running, energised through S_1, D_1 and S_2 which, in the mean time, will have closed. The motors will continue to operate the valves until until by a fluke the two corresponding limit switches open simultaneously. A new signal, however, will set the switching cycle off once more.

With this in mind, most servo motors are fitted with a separate auxiliary switch, which is made to operate the second motor as is shown at (*b*). In two-position control systems, following and sequencing schemes are identical unless in the latter case each valve is operated by its own sensing element.

2.5.4. Day-night control

Whereas in countries with moderate climates it is common practice to switch off the heating at night, considerations of frost protection and the need to limit the heating-up period when starting up again in the morning favour the maintenance of a reduced space temperature in countries with more severe weather conditions. Thus, also, the building is prevented from cooling down too far which, in reducing heat loss due to radiation to the cold walls, from the human body, improves comfort conditions during the early hours of the following day.

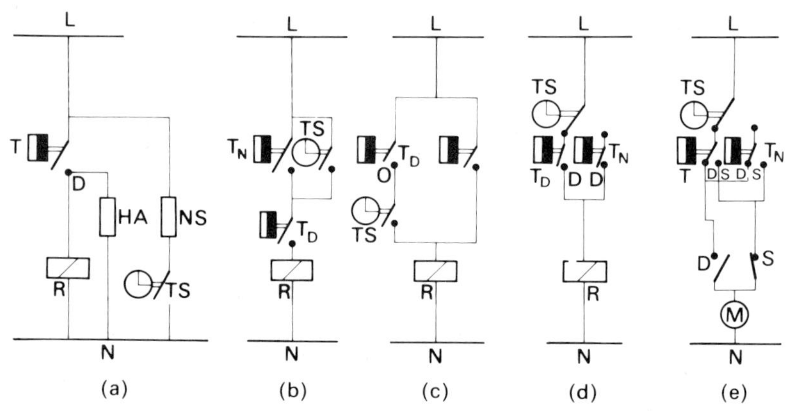

Fig. 2.12
Methods of achieving night set-back; (*a*) Night set-back limited to a fixed value; (*b*),(*c*),(*d*). Alternative methods of connecting up a separately adjustable night thermostat; (*e*) Method to be used when the final control element is to be energised by an s.p.d.t. contact; T_N, T_D, night and day thermostats; TS tuneswitch; OR oil burner relay.

Figure 2.12 shows several methods by which night set-back is achieved. At (*a*) a roomstat is shown to contain an extra heater NS controlled by a time-switch which, closing its contacts at night, causes the temperature-sensitive element to become warmer than ambient and thus to reduce the set value. The temperature drop is limited by the magnitude of the resistance in relation to line voltage and, for practical purposes, is of a given magnitude (e.g.5°C). If a different value is desired, two thermostats are usually placed, one adjusted to the day temperature, the other to the minimum night temperature. At (*b*) and (*c*) the switching over is achieved by an s.p.s.t. contact; at (*d*) and (*e*) an s.p.d.t. contact is used. This, in the case of a motorised valve (*e*), is mandatory.

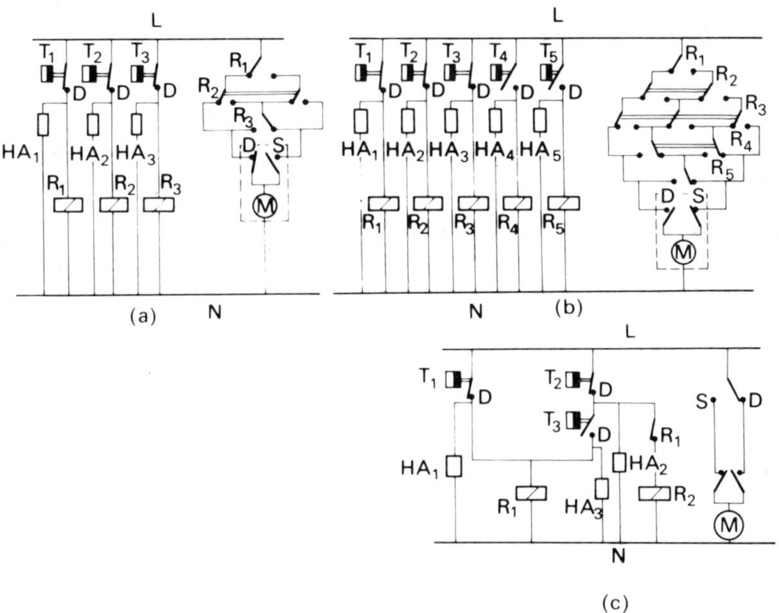

Fig. 2.13
Averaging control in two-positional regulating systems. (*a*), (*b*) correct method in the presence of heat accelerations for three and five detectors respectively; (*c*) incorrect method as the presence of heat-accelerators will put one of the three detectors permanently out of operation; T_1, T_2, etc. thermostats; HA_1, HA_2, etc. heat accelerators.

2.5.5. Averaging control

In large spaces such as auditoria, halls in museums etc, considerable variations in temperature may occur simultaneously. In the early days of automatic control, when two-position control was practically the only method in use, an odd number of sensing elements were placed in strategic positions and it was left to the majority to decide whether heat was to be conveyed to the space or not. Nowadays, air conditioning and more sophisticated methods of control have effectively solved the problem. Therefore, the method is largely redundant, and it will suffice to point out that the link up (*c*) in Fig. 2.3 is wrong if heat acceleration is used, in which case the use of relays as at (*a*) and (*b*) is essential.

2.6. Multi-step control

In Section 1.1 the high/low system was given as an example of a two-step controller which, for long periods it is – or should be. In actual fact it is three- positional as it invariably includes an 'off' position. As such it conforms to the BS standard definition of multi-step controller '. . . whose output signal assumes predetermined values at two or more values of the deviation'. It is useful in cases where, normally, even the least favourable combination of disturbances can not cause the load to fall below a certain value, and where, as a consequence, a minimum load is practically always present. In limiting the control range to a band of values between the maximum and fractionally below the minimum, control quality may be considerably improved owing to the reduction in span.

An installation consisting of two boilers of which one is operated semi-permanently and the other is switched on and off to meet fluctuations in load may serve as an example. Though the load will seldom fall below the capacity of boiler 1, provision must be made for the rare case when it does. This will take the form of an extra high limit device cuts out boiler 1 in an emergency at a value higher than that at which operation is switched from high to low load.

The number of steps can be increased, sometimes considerably, and control quality improved accordingly. However, the greater number of steps, the nearer the scheme approaches modulating control, the subject of the following chapters.

3. Integral controllers and the application of floating control

3.1. General

In contrast to the two-position variety, floating controllers can bring the final control element to a standstill at any point of its stroke and move it from there in either direction as may be required. One version of this group, the *integral controller*, moves its valve at a speed directly proportional to γ ($= \gamma_0 - \gamma_i$) and thus obeys the law

$$\frac{ds}{dt} = -K_2 \gamma$$

or

$$\Delta s = s - s_0 = -K_2 \int_0^t \frac{d\gamma}{dt}$$

in which K_2 is a constant characterising the range of speeds to which the controller is adjusted. In this connection its 'integral action time', T, may be defined as the time taken by the valve to complete the whole of its stroke in response to a sustained deviation of one unit. This response is shown in Fig. 3.1. If γ is reduced to one half of its original value, the speed of the final control element is also halved. From the figure it can be confirmed that when $\gamma = 1$, $-K_2$ equals tan α.

The simplest way to demonstrate integral action is by means of a hydraulic controller shown schematically in Fig. 3.2. A lever, 1, rotating round point 2, is moved up and down by a bimetal spiral. Via A, movable along lever 1 it actuates valve mechanism 5 and 6 in cylinder 7 to which oil under pressure p is admitted through port 4. As the temperature rises, the bimetal curls up and raises point B. Oil flows to cylinder 8 and plunger 9 is moved downwards to close the final control element FCE. At the same time, excess oil is exhausted via 6. The desired value γ_i is adjusted by moving 2 up and down, K_2 by moving A to the left or the right along lever 1.

Pneumatic I controllers (see Chapter 4) are less common. Figure 3.3. shows diagrammatically a pressure controller of which a flapper-nozzle

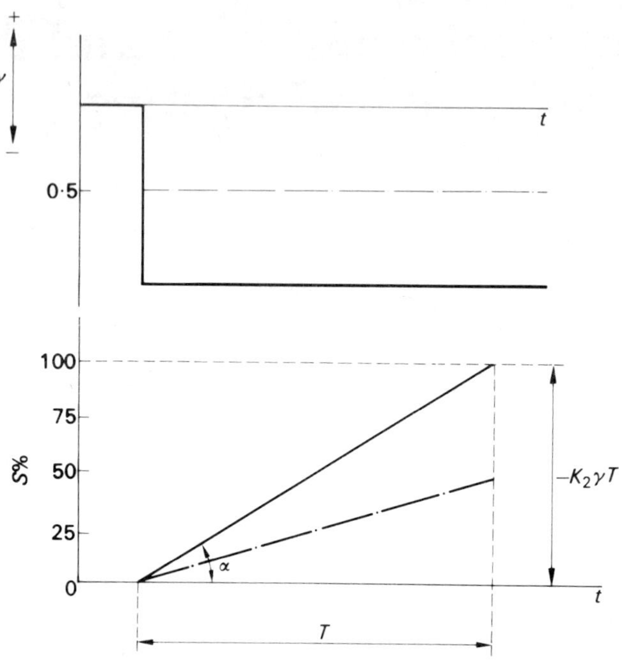

Fig. 3.1
The response of an integral controller to a sustained deviation equal to unity: its actuator travels through the whole of its stroke in a time T, the integral action time.

system, 1, 2, is supplied with compressed air through restriction R_1. As the pressure at 3 rises, diaphragm 4 moves the flapper towards the nozzle 2, and the air pressure in the space between 2 and R_2 increases. This signal is transmitted to the actuator 5 via restriction R_2 followed by a capacity C, and the final control element starts to move. The speed at which the pressure at the motor increases is inversely proportional to the product RC. As C is not easily changed, the integral time is usually adjusted by changing the size of restriction R_2; the desired value is adjusted by means of screw 6. The FCE will continue to move until the pressure at 3 reaches γ_i and will then stop.

True integral control is difficult and rather expensive to achieve by electrical means. Electrical motors with continuously variable speed, and more especially the means of controlling such motors, are too cumbersome and costly to be used on valves or dampers; only squirrel-cage motors running at constant speed are economically feasible for this

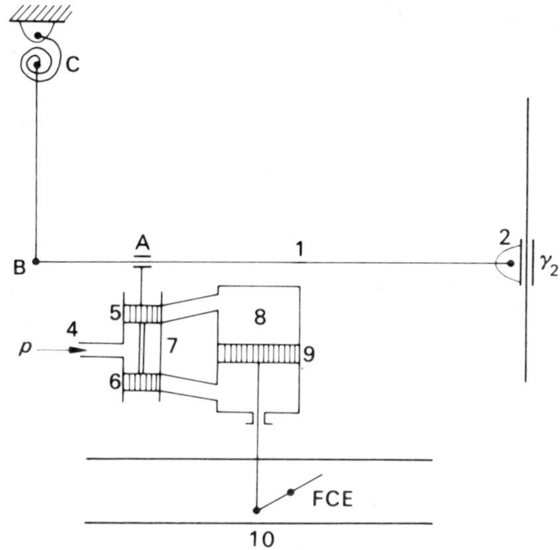

Fig. 3.2
Hydraulic I controller for temperature.

purpose. The adaption necessary has given rise to many different electro-mechanical versions which, while belonging to the family of floating controllers, are no longer truly integral in their action. Figure 3.4 gives further details.

At (a) the action of a pure I controller is shown; the velocity of the FCE is continuously proportional to the deviation γ from the desired value γ_i. As γ is rarely reduced to zero for any length of time, the valve would never come to rest. This would result in a continuous and wholly unnecessary movement and resultant wear. Therefore, as a first step, a neutral zone γ_N is introduced within which γ can assume any value without resulting in movement of the FCE. At (c) the requirement of variable speed is met by supplying the motor with a series of impulses the duration or frequency of which are proportional to γ. Though when the motor moves, it moves with constant speed, its average velocity fulfills the requirements of integral control. A further simplification is shown at (d) where, instead of varying continuously, the average speed increases in a number of discrete steps. As shown, there are 3, the lower speeds being used as long as γ is small. As is the case at (e) where only two steps are shown, the arrangement can be seen as a corresponding number of separate schemes to (f), each speed being allotted its own neutral zone,

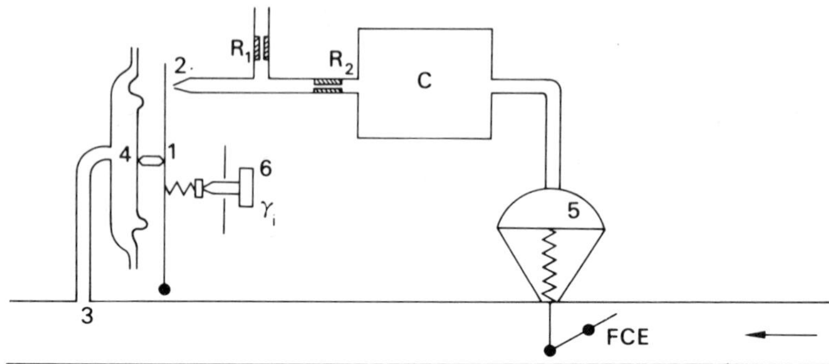

Fig. 3.3
Pneumatic I controller for pressure.

γ_{N1} and γ_{N2} becoming wider as the average speed increases. It is sometimes possible, and occasionally desirable, to adjust the speeds with which the valve opens and closes independently of one another as shown at (*f*). The scheme has advantages where the control constants for the charge and discharge phases differ significantly.

The final arrangement shown is the cheapest of all. It is also occasionally called a floating controller, and consists of an extremely slow-moving motor operated by a detector with a changeover contact. Inasmuch as it never comes to rest it is not a true floating controller. At the same time it is not a two-position controller either, as it can change the direction of its movement at any valve position.

3.2. What these controllers can and cannot do

Each succesive modification means a further step away from the theoretically ideal integral controller. The simplifications are by no means necessarily a step back, it is just that the controllers become progressively different and assume characteristics not present in the pure I controller. Yet they have important features in common: they cannot come to rest until the deviation has been brought within the limits of γ_N.

That this attractive feature so seldom brings corresponding results is due to another inherent characteristic they have in common: their action tends to come too late: they never seem to catch up with events (Fig. 3.5). If a variable has come to rest at the desired value γ_i and is suddenly affected by a disturbance u that drops away soon after as suddenly as

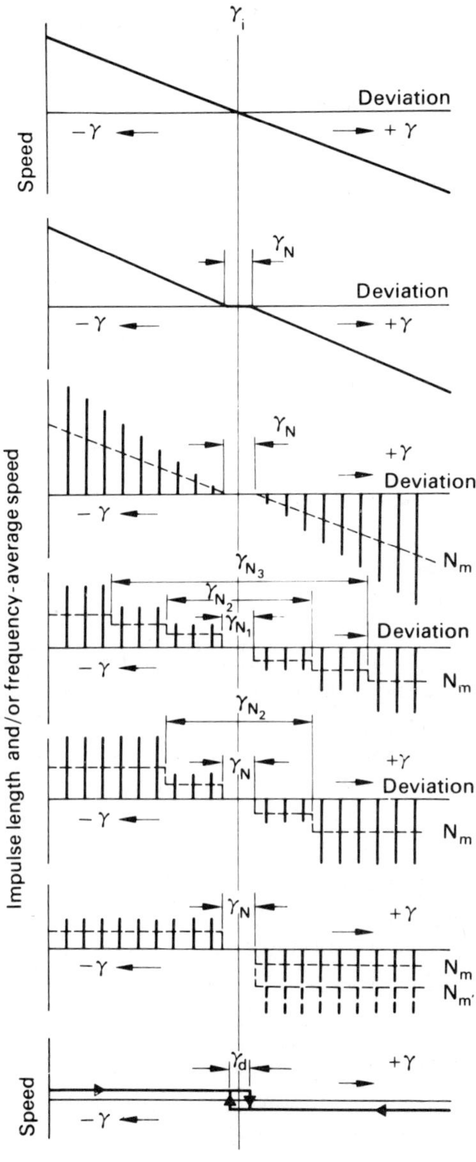

Fig. 3.4
The reaction of various floating controllers energised electrically as a function of the size of deviation γ.

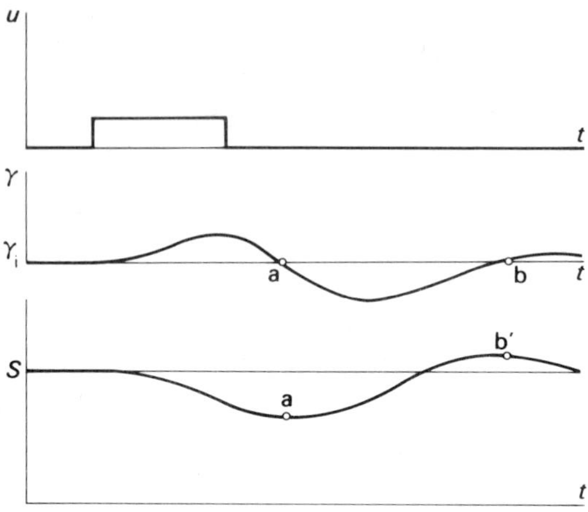

Fig. 3.5
The inherent tendency of floating controllers to lag.

it appeared, the variable will wander away from γ_i. As a result, the valve will start to move, the more quickly the larger γ becomes . When at point *a*, the deviation is reduced to zero as a consequence of the controller action, the valve is in position *a*′, whereas equilibrium in the initial condition indicates that it should have returned to its starting position. A new departure from γ_i results as indicated by the dashed line, referring to a different, but for present purposes similar case. Once more, points *b* and *b*′, show how the valve position lags behind γ. If $\int_0^t (d\gamma/dt)$ gradually drops, the amplitude of the valve movement will decrease in ratio and a new equilibrium be eventually found. Yet it will be clear that if this inherent lag is increased by additional lags present in the control loop, the tendency to oscillate may soon become uncontrollable. This tendency will be the more pronounced the greater the average speed is with which the valve is moved in response to a unit deviation.

3.2.1. A stability diagram

These conclusions are confirmed by the stability diagram (Fig. 3.6) published by Junker[6] for a floating controller to Fig. 3.4(*f*), characterised by the fact that neutral zone, γ_N, impulse frequency, $f = 1/t_0$, and impulse duration it_0 are separately adjustable. The motor, when running continuously needs T_m seconds to complete its stroke. This speed must be considered a given quantity as well as τ, T_{de} and *Sp* that, together, characterise the process (Fig. 3.7). During one impulse

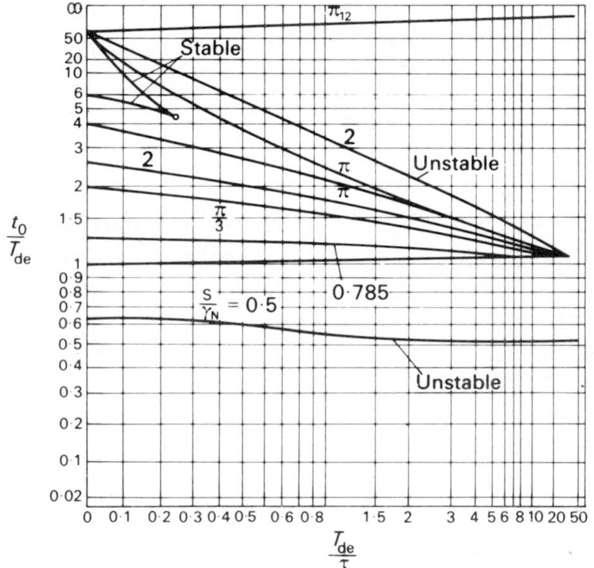

Fig. 3.6
Stability of a controller to Fig. 3.4 (*f*) as a function of the control characteristics of the process controlled and the adjustment of the controller according to Junker.[6]

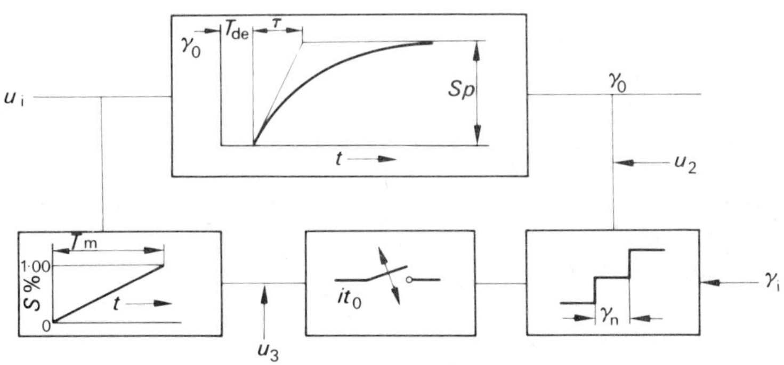

Fig. 3.7
Block diagram of a control loop in which a simple floating controller is used.

it_0 the motor moves over $(it_0/T_M) \times 100$ per cent of its stroke and causes a potential step change S equalling

$$S = \frac{it_0}{T_M} 100\% \times a_p = \frac{it_0}{T_M} Sp$$

expressed in units of the controlled variable (for a_p see Section 1.6.2). If S is very large with respect to γ_N, the controlled variable will keep on skipping from one side of $\gamma_i \pm \gamma_N/2$ to the other without ever coming to rest within the neutral zone. On the other hand, the diagram indicates that, if S/γ_N is made too small, this can also result in instability unless t_0 is increased with respect to the dead time T_{de} . In other words, Junker's graph indicates the value of S/γ_N for which, at given values of t_0/T_{de} and T_{de}/τ, instability resolves into continuous action and vice versa; the S/γ_N curves in the graph are therefore stability limits.

If t_0 is chosen equal to T_{de}, the graph shows that S/γ_N will have to be larger than 0·785 if stability is to be maintained – and this value is wholly unaffected by T_{de}/τ. Adjustment to this value, however, would result in a fairly lazy control system, unable to keep pace with deviations materialising swiftly. Increasing S/γ_N however, may mean reducing the impulse frequency, the ratio T_d/τ assuming more importance as the larger values for S/γ_N are chosen. At $S/\gamma_N > \pi/2$ an upper limit to stable operation appears and, as the value of this parameter increases, stability is progressively reduced to very small values of T_{de}/τ.

To take a concrete example, Heck,[5] who carried out extensive measurements on an air heater, found that, in one case,

$$\frac{T_{de}}{\tau} = \frac{92\text{s}}{268\text{s}} \quad \text{or} \quad 0{\cdot}342$$

Referring to the graph this plant appears to be very suitable for the application of the controller of this type as S/γ_N can be made equal to π provided $2{\cdot}8 < t_0/T_d < 4{\cdot}6$, or, with T_{de} at 92s, $258\text{s} < t_0 < 450\text{s}$. In general it will prove preferable to choose a smaller ratio of S/γ_N and to increase the impulse frequency, but how far one should go depends on the circumstances.

3.2.2. Adjustment by trial and error

Usually neither τ or T_{de} are known, if no time is available to establish their value by measurement, this found, adjustment has to be made empirically. The average speed of the valve is increased until, for example, any change in γ_i in instability. The valve is then slowed down to

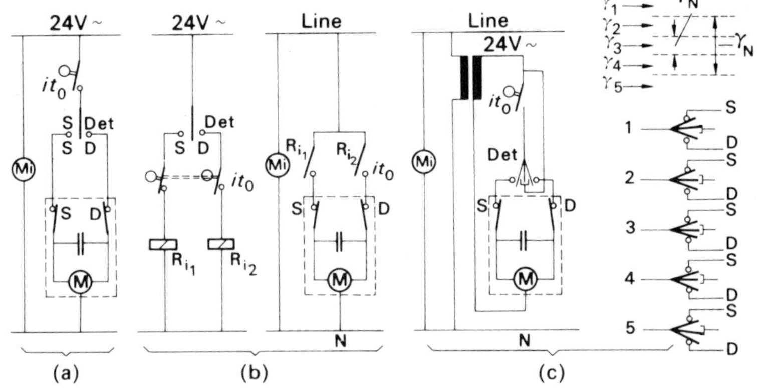

Fig. 3.8
Some simple electro-mechanical floating controllers (*a*) with a single, if adjustable, mean valve speed; (*b*) as (*a*) for line voltage; (*c*) a version with two main valve speeds, one of which is adjustable.

50–80 per cent of the critical value, reducing the margin of safety to the extent that the conditions under which the adjustment was made were less favourable.

3.3. Some examples of electrical floating controllers

It is usual to arrange the circuit for low voltage as the sensing element contacts seldom have snap action (Fig. 3.8). Even so they are sensitive to vibration. In this respect the it_0 contact which normally has snap action fulfills a useful secondary function in reducing sensing element contact wear. When the motor, of which the contacts limiting the stroke are both closed in all but the two extreme positions, is suitable for mains voltage, it must be energised via relays R_{i1} and R_{i2} (Fig. 3.8 (*b*)). At (*c*) a controller as in Fig. 3.4 (*e*) is shown. It has two speeds and two neutral zones γ_{N_1} and γ'_{N_2}. As long as the deviation is small, the motor will be energised via the impulse contact it_0 as the detector contact will assume one of the positions 2 or 4. If the deviation grows to a value $> \gamma_{N\frac{1}{2}}$, the contact will assume either position 1 or 5; the impulse contact is bridged and the motor moves continuously. Chopper-bar controllers (Fig. 3.9) are among the most common industrial examples of this group; these, also, can be supplied for more than one mean valve speed.

Figure 3.10 shows an 'electronic' type. After amplification, the out-

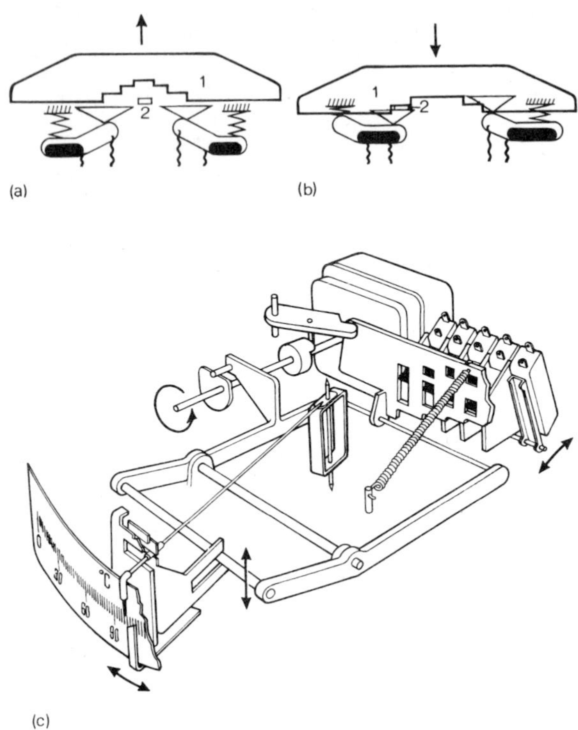

Fig. 3.9
Typical chopper-bar controller with three mean valve speeds. The duration of the impulse it_0 depends on the magnitude of the deviation γ. 1, Chopper-bar, 2, Switch operator actuated by chopper-bar which moves to the left or to the right in sympathy with sign and magnitude of γ and is thus depressed for a shorter or longer time.

put signal from the Wheatstone bridge operates an output relay which energises resistances R_4 or R_5 placed in parallel with the respective motor coils. The heat thus generated changes the resistance of thermistors R_2 and R_3 and thus re-balances the bridge circuit before γ has wholly disappeared. The motor stops and the heat supply is discontinued; while R_2 or R_3 cool down, the change in valve position affects the controlled condition, further impulses will ensue as long as the reduced input signal

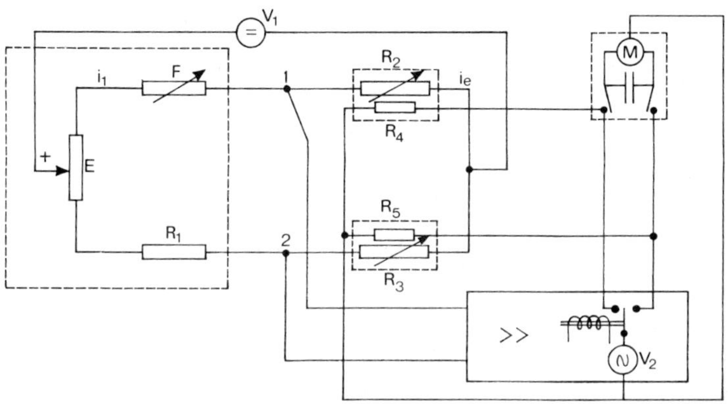

Fig. 3.10
Electronic version of a floating controller

exceeds γ_N, which, in this case, is determined by the sensitivity of the relay. The larger the deviation, the larger also must the feedback signal be to re-balance the bridge and the longer the motor must run to produce the necessary heat. This also implies a higher R_2 or R_3 temperature and therefore a more rapid rate of cooling. The scheme approximately fulfills one of the requirements of a truly integral controller. If adjustability of the I time is desired the thermal feed-back circuit is moved to a separate bridge, in series with that of the sensing element.

3.4. Practical application of controllers of the integral type

3.4.1. Main fields of application

As has been shown, controllers belonging to this group are very sensitive to lags and relatively indifferent to the value of τ. They are at their best when deviations develop slowly. As far as the heating and air-conditioning field is concerned there are four major applications for floating controllers: that of flow temperature after a three-way valve; or after a non-storage calorifier; of dew-point; and of supply air temperature. Furthermore, they are particularly suitable for the control of flow rate and – in many cases – for pressure control applications, owing to the negligible transfer – and velocity-distance lags.

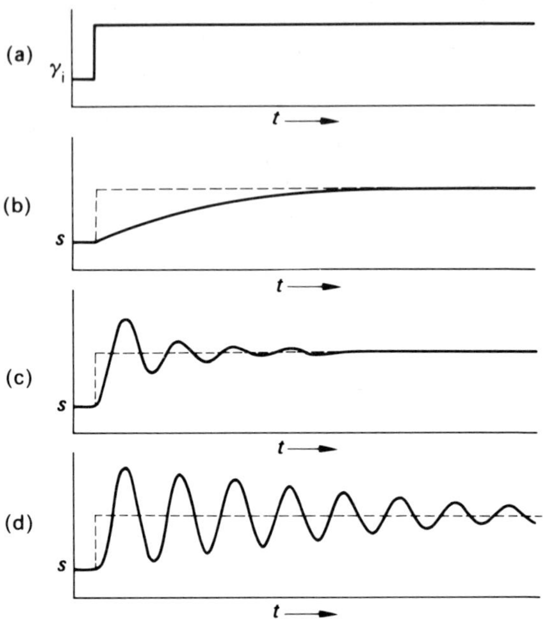

Fig. 3.11
The response to a step change (a). At (b) the integral action time is too long; at (d) too short; and at (c) about right.

3.4.2. Adjustment

Figure 3.11 shows three types of response to step change (a). Aperiodic response (b) is usually too sluggish; especially when the deviation develops rapidly, it is not corrected quickly enough. On the other hand, if the control action is too vigorous, the limit of stability may be approached (d) or exceeded. It is therefore advisable to aim at a reaction similar to (c) some latitude in either direction being permissible.

The adjustments possible on commissioning the installation vary with make and type of the controller. The effect that the parameters have was shown in detail in Fig. 3.7. As a very rough guide, the control action gains in stability with increasing t_0, γ_N and T_M and with decreasing i and Sp. The latter can usually only be reduced by pre-control of, for example, the flow temperature in heating applications.

3.4.3. Limit control

If a two-step limit is acceptable, Fig. 3.12 (a) for low limit and (b) for high limit show arrangements familiar from Fig. 2.10. Generally it is

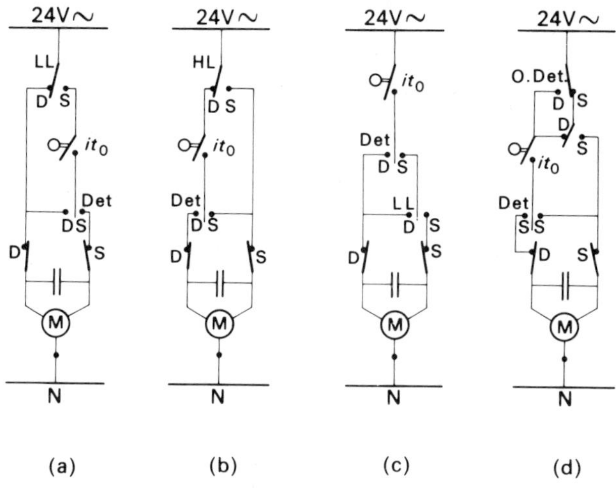

Fig. 3.12
Limiting arrangements at (*a*) and (*b*) are correct, at (*c*) incorrect.
The arrangement at (*d*) reduces the effective span at low load.

important that the limiting detector functions independently of the impulse contact. For this reason the scheme show at (*c*) is inadvisable. It illustrates a further shortcoming inasmuch as the contact of the actual detector, assuming its mid-position, can prevent the limit stat from functioning. However, there are applications where the limiting action should be prevented from being too vigorous (*c*), in which case the impulse contact is placed before the limit stat. Scheme (*d*) limits the overshoot by reducing the flow temperature at low load; its contact being short-circuited when an outside detector switches over at lower outside temperatures.

These schemes all provide a rigid limitation. Electro-mechanical controllers of the I type do not lend themselves readily to flexible limitation; electronically this is simple enough (see Chapter 5).

3.4.4. Sequencing two final control elements

The amplification factor, and thereby the size of each step *S*, can be reduced by sequencing two valves, in which case both motors may be fitted with auxiliary contacts (Fig. 3.13 (*a*)). When M_1 reaches the end of its stroke its D-contact opens and auxiliary contact AC_1 closes, thus transferring the command to the D-terminal of M_2. At the same time

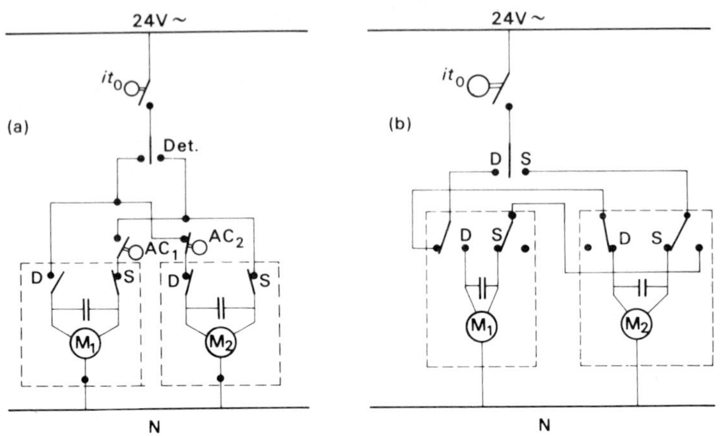

Fig. 3.13
Two electrical actuators operated in sequence by means of (*a*) auxiliary, (*b*) built-in switches. In both cases M_1 has opened its valve fully, whereas M_2 is being modulated.

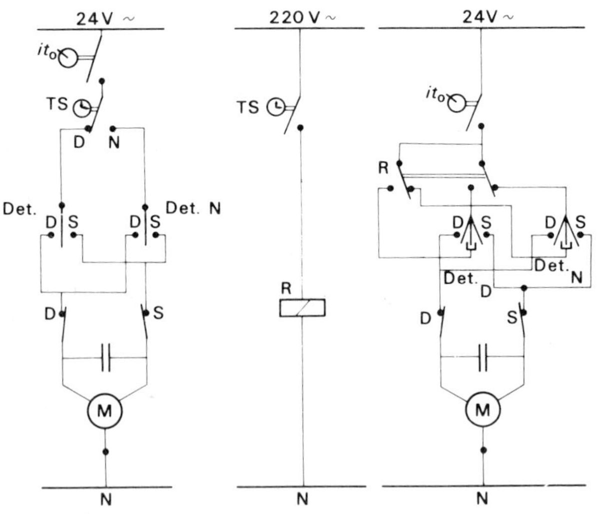

Fig. 3.14
Day-night control in simple electro-mechanical floating control applications

AC_2 opens so as to prevent all movement of M_1. When the load decreases once more, M_2 must close its valve so that its contact S opens and AC_2 closes before the command signal is returned to M_1 and it can start to close in its turn.

At (*b*) both motors are fitted with s.p.d.t. stroke-limiting contacts in which case no auxiliary switches are required. In air-conditioning plants the heating and cooling valves are frequently sequenced. As the *Sp* for the heating phase is usually a factor two or more larger than for the cooling phase, this would mean that the amplification factors differ in the same ratio, which is undersirable. As it is not feasible to change it_0 or the integral time at the moment of switching over, the solution is usually sought in the linkage between motor and valve. If a slave motor is required to move its final control element in sympathy with the master motor ('following control'), a balancing relay of a type to be described in the next chapter can provide the necessary flexibility.

3.4.5. Day-night control

In simple electro-mechanical schemes two detectors are provided, and the time switch energises either one or the other, as required by means of an s.p.d.t. contact (Fig. 3.14 (*a*)). With two-speed controllers the time switch must be fitted with a d.p.d.t. contact or a double-pole relay be used (*b*).

4. Proportional controllers

4.1. General

Owing to their simplicity, adaptability and stability, proportional controllers are the most widely used type of modulating controllers. Unlike those of the integral family where any valve positon can occur at any value of the controlled variable, there is a firm relationship between valve position, s, and deviation, γ: the proportionality of these two magnitudes has given this group its name:

$$\Delta s = -K_1 \gamma$$

K_1, the proportional action factor, is ideally a constant, characteristic for the adjustment of the controller. The set value need not be related to the valve mid-position as in Fig. 4.1 (*b*); the choice of $s = 0$ or 100 per cent is nearly as common for the purpose but any other value will do equally well. Whichever relationship between γ_i and s is chosen, one of the main characteristics of this type of controller is obvious: when γ_0 equals γ_i the valve can only assume one distinct position. If the load is such that it could best be matched at a different value of s, this alternative position can only be taken up if a departure from γ_i is allowed or if the set point is changed.

4.1.1. Proportional band and offset

As pneumatic controllers in their simplest form are by nature of the proportional type and need adaptation if their mode has to be changed, it is fitting that they should provide the model on which to demonstrate the main characteristics of the P type of regulator.

The flapper-nozzle system of the pneumatic temperature controller of Fig. 4.2 generates a signal between 0·2 and 1 bar (3 to 15 p.s.i.) by relating distance d, and therefore pressure P to γ_0^0. Ideally P is a linear function of γ: the ratio between the two can be adjusted by, for example, sliding the point at which the movement of the detector is transferred to the flapper to the left or right. The signal is transmitted

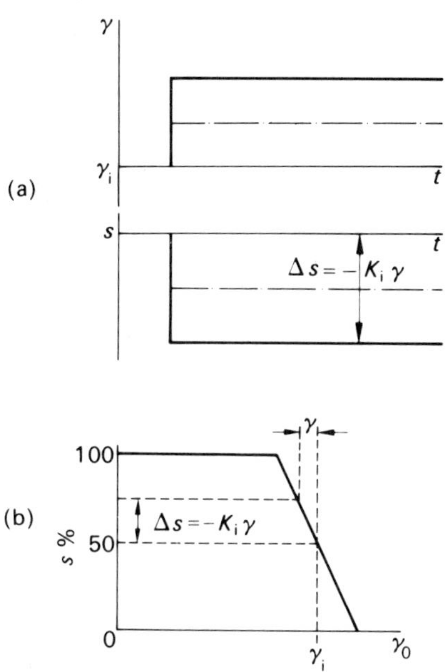

Fig. 4.1
Proportional control action: Two ways of showing the relationship between the deviation, γ, from the set value, γ_i, and the resultant displacement of the final control element, Δs.

to the pneumatic actuator shown mounted on a valve which, as illustrated, closes when the signal drops to its minimum value. As the signal increases, the valve disc is forced open against the growing spring pressure that resists the movement. If the variable spring pressure were replaced by a constant force, e.g. by a weight acting via a lever, the valve would continue to open until the temperature had been restored to the set value by the increase in flow through the valve; the controller would then be of the integral type. It is the change in spring tension with the valve movement that prevents the valve from travelling as far as it otherwise would; it is this that makes the regulator a proportional controller. From this example the following conclusions may be drawn:

(*a*) that, as long as the controller is in a position actively to resist undesired changes in γ_0, γ will vary between the values at which the valve is either wholly closed or fully open. The difference between the two expressed either in units of the controlled variable is called the

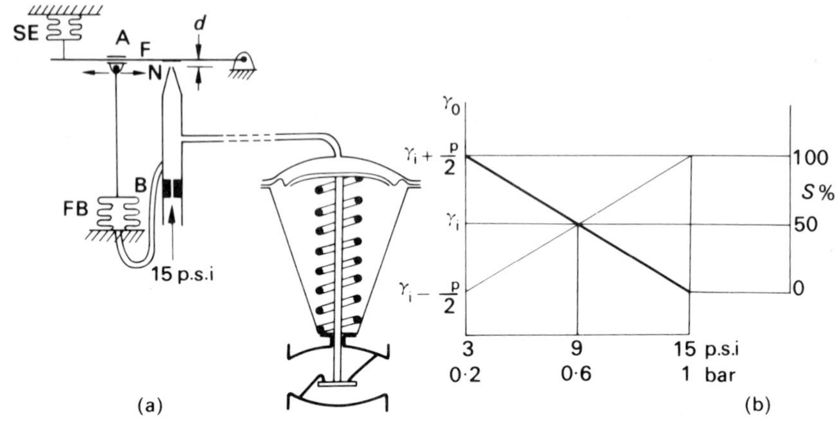

Fig. 4.2
(*a*) schematic arrangement of a pneumatic P controller of the 'bleed' type SE, sensitive element; FB, feed-back bellows; F, flapper; N, nozzle (*b*) the idealised relationship between controlled variable γ_0 on the one hand, output signal, *p*, and valve position, *s*, on the other.

throttling range, or proportional band, γ_P of the controller. (Fig. 4.2(*b*));

(*b*) that P controllers tend to reduce deviation rather than eliminate it; the remaining, sustained, deviation being called the offset or, in the United States, droop;

(*c*) that only when the load corresponds to the position the valve assumes when $\gamma_0 = \gamma_i$ will the offset equal zero;

(*d*) that in the above arrangement the fully open valve position cannot be attained unless an offset is allowed. This means that, even if the valve capacity were adequate totally to remove all deviation, its ability to do so would be artificially limited by the nature of this mode of control;

(*e*) proportional band and offset are co-related: halving the one halves the other; from this point of view the best results are achieved at the smallest possible γ_P.

A second definition of the proportional band, commonly used in industry, relates the input to the output signal, both expressed as a percentage of the control range. This ratio, itself also expressed as a percentage is then used as a measure of γ_P. If in a given process ($Sp = 20°C$) a deviation of 2°C (10% of *Sp*) causes γ_0 to change potentially by 10°C or 50% of *Sp*, $\gamma_P = (10/50) \times 100\% = 20\%$ (Fig. 4.3). At

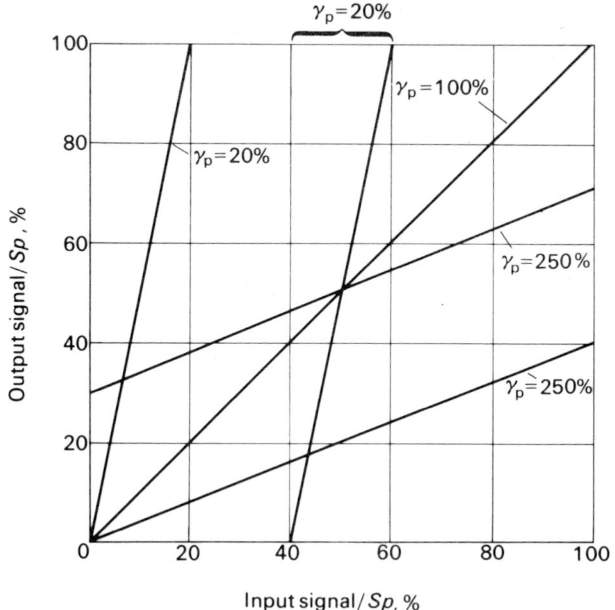

Fig. 4.3
The proportional band expressed as a percentage of the control range of the process.

γ_P = 100 per cent, the potential sustained deviation may increase to $Sp/2$ and at 200 per cent only one half of the control range will be used.

4.2. Feedback and amplification factor in P control

If there is to be a firm relationship between valve position and deviation, there must be a signal proportional to the change in s with which the signal proportional to γ can be compared. This arrangement is called the feed-back of the controller, as a signal originated at a later stage is transmitted to an earlier stage. The feedback signal is called negative as it is made to oppose the signal to which it owes its existence. Its effect is therefore essentially to limit the valve movement thereby increasing the stability controllers of the integral family lack due to the absence of such constraint. With pneumatic controllers, the feedback signal was shown to be the change in spring tension which is compared with the signal applied to the actuator. The feed-back signal is usually increased by means of a

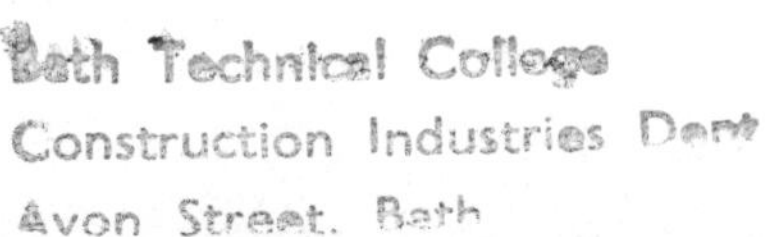

bellows to which the output signal is applied and which, in expanding and retracting, exerts a force on the flapper opposed to that of the sensitive element. Thus the flapper movement is restricted and the output signal less than it would have been without feed-back bellows. In electromechanical and electronic versions it generally takes the form of a potentiometer, the rider of which is moved by the motor spindle. In hydraulic and many simple types of mechanical controllers, the detector initiates a movement proportional to γ with which the final control element is made to move in sympathy.

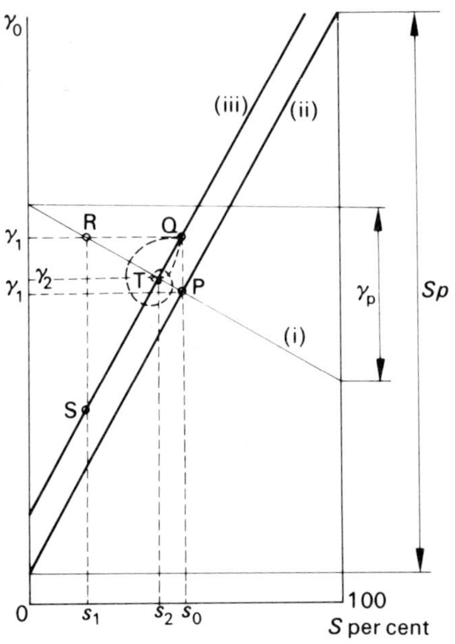

Fig. 4.4
Proportional (I), and process control characteristics (ii) and (iii) superimposed.

It now has to be decided what the optimum ratio is between the two signals, defined as K_1 in Section 4.1. To this end Fig. 4.1 (*b*) is superimposed upon Figures such as 1.29 or 1.30. Ideally, both characteristics are straight lines (Fig. 4.4).

When the controller has brought the valve to rest in a position corresponding to the value of the variable that gave rise to the valve movement, the system is in balance. In Fig. 4.4 this is the case at point *P* on controller characteristic (i) corresponding to $\gamma_0 = \gamma_i$ on the ordinate, and valve position s_0 on the abscissa. P.C.C. (ii) shows that, at s_0, the variable potentially equals γ_i under the given load conditions, so that the loop is closed and balance achieved. Now assume that, due to a step change, p.c.c. (ii) suddenly shifts to p.c.c. (iii). The controlled variable will rise to Q. This new value corresponds to γ_1 and to point R on the controller characteristic, and the valve will start moving towards position s_1. In the absence of any lags whatsoever, the controlled variable will respond by moving downward along p.c.c. (iii) and come to rest at point T, the point of intersection between lines (i) and (iii). In the presence of lags, however, there will be a tendency to overshoot. Thus, at s_1 γ_0 will potentially drop to point S which is far too low. Even though neither point is likely to be reached, the continuous overshoot will keep on causing the direction of the valve movement to reverse until, ultimately, balance is once more achieved at s_2 with the controlled variable at γ_2. The locus of the points showing the values of γ_0 and of s that have occurred simultaneously may look something like the dashed spiral that ends up in the new point of equilibrium, T. It is seen that the potential deviation $\gamma_1 - \gamma_i$ has been reduced to the offset $\gamma_2 - \gamma_i$.

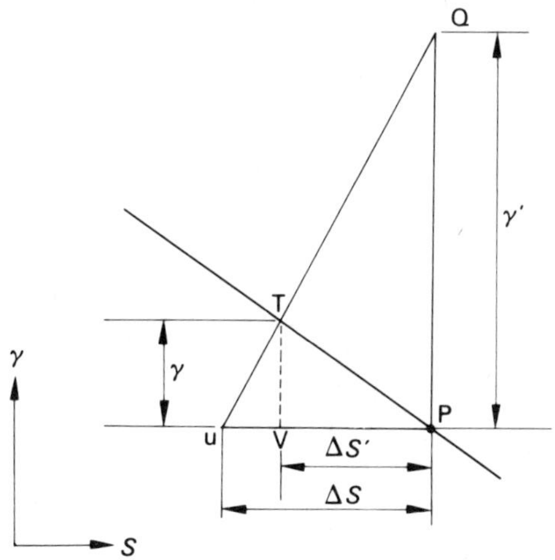

Fig. 4.5
Potential and actual deviations γ' and γ in a P controlled process.

4.2.1. The gain in a proportionally controlled process

Figure 4.5 shows a section of Fig. 4.4 on a larger scale. Triangles UPQ and UVT are similar;

therefore $$\frac{\mathrm{TV}}{\mathrm{QP}} = \frac{\mathrm{UV}}{\mathrm{UP}} \quad \text{or} \quad \frac{\gamma}{\gamma'} = R = \frac{\Delta s - \Delta s'}{\Delta s}$$

in which R is the factor by which the controller has reduced the potential deviation γ' to the offset, γ.

In accordance with the definition of a P controller

$$\Delta s' = -K_1 \gamma,$$

whereas section 1.6.2 showed that

$$a_p = \frac{\gamma'}{\Delta s}, \quad \text{or} \quad \Delta s = \frac{\gamma'}{a_p}.$$

Inserting these values in the expression for R, one obtains:

$$R = \frac{\dfrac{\gamma'}{a_p} - K_1 \gamma}{\dfrac{\gamma'}{a_p}} = 1 - K_1 a_p \frac{\gamma}{\gamma'} = 1 - K_1 a_p R$$

Thus,

$$R(1 + K_1 a_p) = 1, \quad \text{or} \quad R = \frac{\gamma}{\gamma'} = \frac{1}{1 + K_1 a_p}$$

If the product $K_1 a_p$ is equated to the gain, G, of the loop controlled proportionally, we obtain

$$R = \frac{1}{1 + G}$$

In the ideal case shown in Fig. 4.3,

$$a_p = \frac{Sp}{s}, \quad \text{and} \quad K_1 = \frac{s}{\gamma_p}$$

so that

$$R = \frac{1}{1 + \dfrac{SpS}{S\gamma_p}} = \frac{\gamma_p}{\gamma_p + Sp}$$

and when

$$Sp \gg \gamma_p, \quad R = \frac{\gamma_p}{Sp}.$$

In reality neither characteristics are necessarily, or even likely to be, straight. The p.c.c. may be considerably more curved than is shown in Figs. 1.29 and 1.30, and the P controller characteristic may assume a shape like line (*a*) in Fig. 4.6 for a pneumatic controller. Since for small changes in γ_0 and s the curvature can be neglected, Fig. 4.4 remains applicable. Thus it will be seen that a_p, K_1 and, therefore, G are liable to change with valve position. Furthermore, as both a_p and K_1 are liable to be greater than average at small values of s, so that when the valve starts to open, G is likely to be particularly large. It will be shown that this may well lead to instability, one of the reasons why many processes are difficult to control at low load and become more stable as the load increases. This is frequently the case in heating and ventilating applications, so that if the circuit is adjusted to give satisfactory results under these conditions, it is likely to remain stable when, at higher loads, the valve has to open up further.

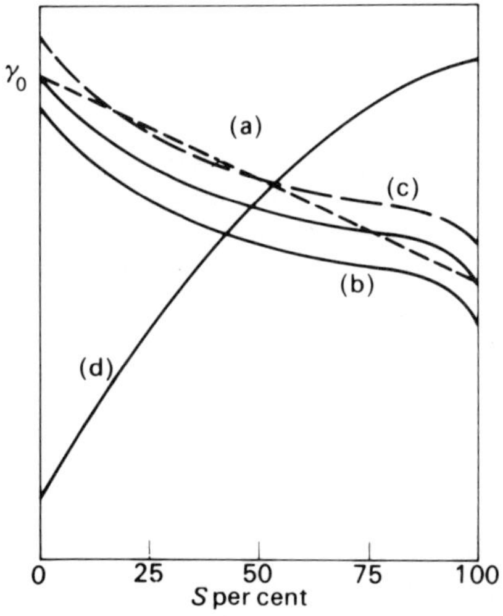

Fig. 4.6
Lack of linearity and of sensitivity in P control changes the controller characteristic into an area. (*a*) notional P characteristic; (*b*), (*c*), limits within which the control point may lie; (*d*) typical shape of a p.c.c.

Figure 4.6 also shows the effect of friction in the valve gland, play in the linkage between motor and valve, lack of sensitivity of the output relay, etc. Due to these and similar effects a deviation must reach a certain magnitude before it is corrected. Thus the characteristic line becomes an area and the controller develops hysteresis. Theory therefore shows that:

(*a*) the narrower the proportional band is made and the larger therefore K_1, the smaller the offset is likely to be;

(*b*) the smaller *Sp* is, the smaller a_p other conditions remaining the same. In view of the fact that in any given case $G = K_1 a_p$, a P controller is likely to give better results, the smaller *Sp* can be made;

(*c*) here, once more, an open loop element reducing *Sp* has a favourable effect on stability when a P controller is used;

(*d*) if a process is controlled by two valves in cascade as, for example the heating and cooling stages in air conditioning, where the spans are unequal, full advantage of this fact should be taken by reducing γ_p of for example the cooling stage of which *Sp* is likely to be the smaller so as to equalise the gain throughout the process.

4.3. The influence of the control 'constants' τ and T_{de}

So far, no mention has been made of the process characteristics other than the control range. The reader may be under the impression that large values of *G* lead to instability under all circumstances; this is incorrect. In the absence of velocity-distance or transfer lags, a P controller would respond aperiodically, and the deviation occurring as a result of a disturbance would be smaller the longer the exponential lag of the process. Effective dead time, however, may cause this stable system to oscillate. Assume that it is of a purely velocity-distance nature, T_d, that the controller is adjusted critically, i.e. with a fairly narrow γ_p, and that the sensing element and actuator are so fast that the valve movement is practically synchronous with γ_0. Under these circumstances the condition shown schematically in Fig. 4.7 may arise: Due to a temporary disturbance, γ_0 may rise to γ_1 and the valve move from s_0 to s_1 as a result. After the lapse of T_d the change in valve movement will make itself felt. γ_0 will start to drop and the valve to move in sympathy. Though a change in *s* of a few per cent would have sufficed to counteract the deviation, the narrow γ_p may have caused the valve to move over several times that distance. Therefore the controller itself causes a new deviation with reversed sign here assumed to be equal to the original deviation. The valve, following γ_0, will once more move too far in the

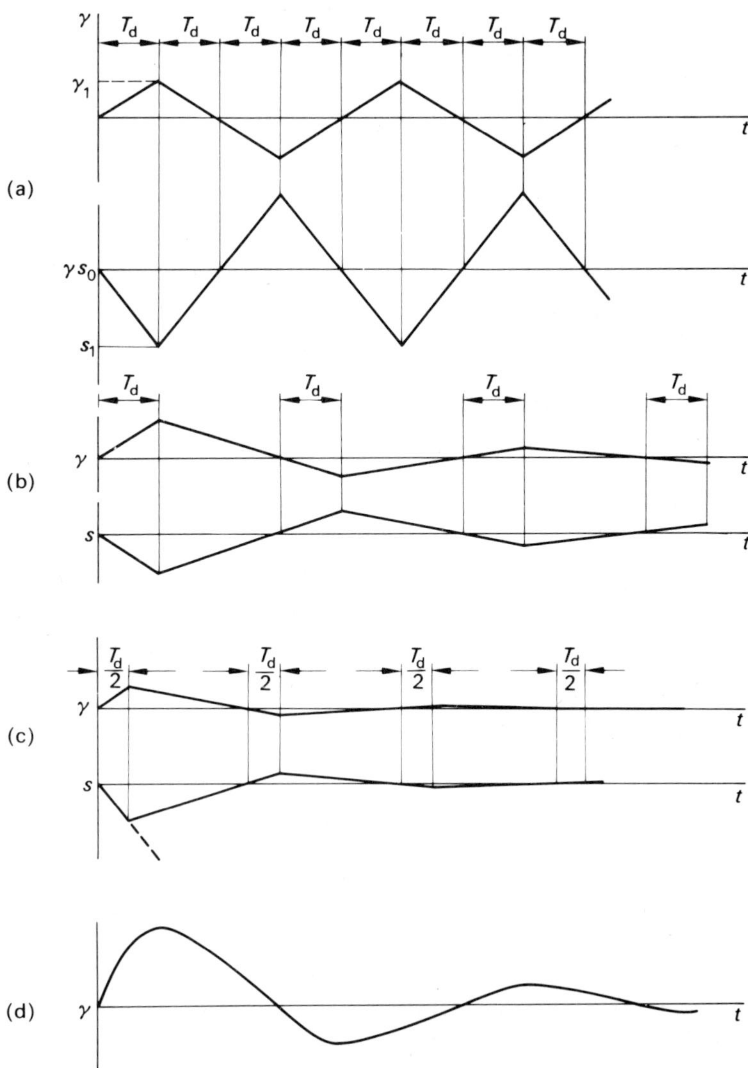

Fig. 4.7
Schematic presentation of (*a*), a P controller adjusted critically; (*b*) an increase of γ_p or of τ as well as a reduction of *Sp*, or (*e*) of T_d increases stability; (*d*), actual shape of a $\gamma_0 = f(t)$ curve in the presence of T_{de} and τ.

opposite direction and so on: the system will continue to oscillate with constant amplitude and without, of itself, being able to come to rest.

Had γ_P been just a little narrower, the valve movement would have gradually increased until the final control element travelled backwards and forwards through the whole of its stroke. Had it been a trifle wider, the amplitude of the oscillation would gradually have diminished and eventually a new balanced condition found (Fig. 4.7 (*b*)). Reducing the span or increasing the exponential lag of the process have a very similar effect, and reducing T_{de} causes equilibrium to be attained more quickly still (*c*). It will be obvious that in practice the lines will be curved rather than straight (*d*), due to the non-linearity introduced by the various lags.

4.3.1. Controller inertia and insensitivity added

Actually, no controller in existence follows fluctuations in the controlled variable without delay. Whether the insensitivity is mainly introduced by the sensing element as in most direct-acting controllers, by the final control element, as in some electric controllers, or by both as may be the case with pneumatic versions, the regulating system can usually be assigned a time constant τ_c with which to describe its reaction to changes in the controlled variable. As pointed out in Section 1.4.1, this is not necessarily a bad thing. Providing the disturbances affect γ slowly, the stabilising effect of τ_c may permit smaller values of γ_P. Yet, if control quality is to be optimum it should be borne in mind that, the process characteristics permitting, controller inertia should be kept to a minimum at all times. The fact that it stabilises the system should not lead to the conclusion that the controlled variable is kept at a constant value, but rather that γ moves too fast for the controller to follow.

Controller insensitivity as illustrated in Fig. 4.6 may also have a stabilising effect when kept within bounds, but for larger values of G and larger ratios of T_{de}/τ, friction etc. may itself cause instability.

Junker[7] confirms by means of a rough calculation that the effect of controller friction and insensitivity is more pronounced the smaller the ratio T_{de}/τ and the larger G. Furthermore, that for each value of T_{de}/τ there is a degree of insensitivity that causes instability at the lowest value of G. The stability diagram of Fig. 4.8, taken from the same publication was worked out for this least favourable condition.

Let for example a very fast P controller ($\tau_e/T_{de} \to 0$) be applied to the air heater examined in Section 3.2.1 for a controller of the integral type. There, T_{de}/τ equalled 92s/268s = 0·342. The graph shows that G

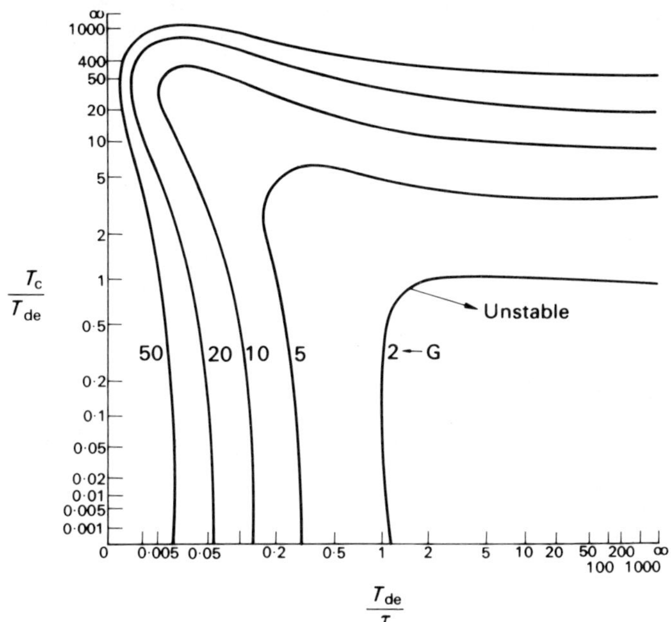

Fig. 4.8
Stability limits for P controllers with inertia and a degree of insensitivity, τ, τ_c time constants of process and controller respectively; T_{de} effective dead time, G gain.

could not be more than 4 without instability occurring. At a span of 50°C that would mean a γ_P of no less than 12·5°C – far too large a value. Had τ_c been of the order of 200s, G would have to have been less still for stability. τ_c would have to be increased by a further factor 10 in order appreciably to reduce γ_P. On the other hand, if τ had been a factor 5 larger, a fast P controller could be adjusted to $G > 10$ ($\gamma_P \sim 5$°C). In general it can be said that a P controller will give reasonable results when $\tau > T_{de}$, Sp is not too large and τ_c is either very short or, when γ changes very slowly, large with respect to T_{de}.

4.4. Different kinds of P controllers

Figure 4.9 illustrates various types of P controllers. At (*a*) a typical self-acting temperature controller is shown. The sensing element (*a*) is

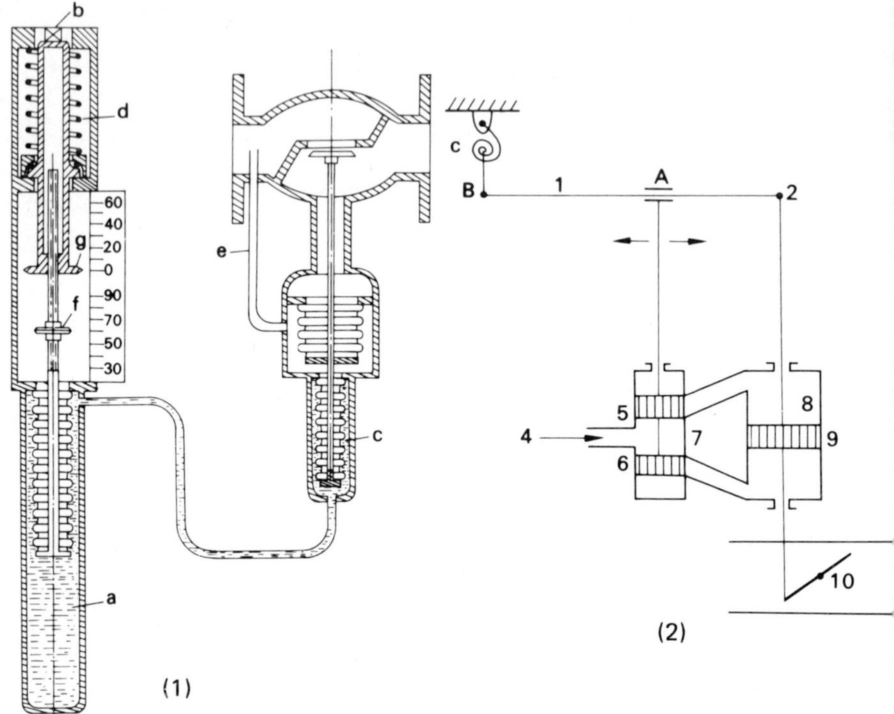

Fig. 4.9
Types of P controllers. (1) A self-acting temperature controller. (*a*), sensing element; (*b*) adjustment; (*c*) actuating bellows; (*d*) overload protection; (*e*) balancing pipe for differential pressure; (*f*) indicator; (*g*) excess temperature indicator. (2), a hydraulic temperature controller.

filled with a fluid which expands with rising temperature. The increased pressure is transmitted to bellows (*c*) which are compressed and cause the valve to close. Should the temperature continue to rise, spring (*d*) will be compressed and the temperature rise above the set value is indicated on the upper scale. In view of the limited force available to close or open up the valve against the differential pressure, the valve is generally balanced as shown at (*e*) or a double-seated version is chosen. The hydraulic P controller (*b*) is reminiscent of the one shown in Fig. 3.2, but the fulcrum 2 is now attached to the plunger 9 which moving in accordance with the position of the servo-valve system 5, 6 and 7, tends to return plunger 9 to its mid-position when a temperature change at C has raised or lowered arm 1. γ_P can be adjusted by positioning point A; the desired value by adjusting the length of *BC*. A pneumatic P controller was shown in Fig. 4.2. The feedback is provided by the sping in the valve motor; its strength or weakness determines γ_P which is further adjustable by positioning point *A*; γ_i is changed by adjusting the length of *AB*.

Most electro-mechanical and electronic controllers are based on a bridge circuit (Fig. 4.10) which frequently consists of two sets of series resistances in parallel (*a*). In the simplest case all four resistances are equal, but provided *a* equals *c* and *b* equals *d*, or even as long as $a:c = b:d$ a voltmeter placed between points A and B will not react when the bridge is energised by an a.c. or d.c. voltage as shown.

If, however, one of the resistances is made to change with the controlled variable (*b*), a voltage will appear between A and B, at any but one clearly defined value of γ_0. By a suitable choice of the resistances this value can be made to equal γ_i, in which case any signal, i.e. any voltage developing between A and B, indicates that the controlled variable has departed from the desired value. After amplification, this signal can then be used to change the position of the final control element. The latter, in turn, can be made to affect the corresponding resistance *d* and thus re-establish equilibrium at a new valve position and a new value of γ_0.

A common type of electro-mechanical controller is built around a similar circuit in which resistances *a* and *c*, as well as *b* and *d* are combined to form single potentiometers, the sliders of which are operated by the detector and the valve motor respectively. Furthermore, two equal coils C_1 and C_2 are incorporated in the bridge circuit (Fig. 4.10 (*c*)); together they operate a balancing relay BR the contacts of which energise valve motor M. As the temperature rises, bellows D expand and resistance *a* is reduced with respect to *c*. The current through the branch containing C_1 becomes larger than that through the branch of which C_2 is a part. The contacts of balancing relay BR tip over in an anti-clockwise direction, energising terminal S of motor M which, the limit switch permitting, starts to close the valve. Attached to the motor is the second potentiometer, made up of the original resistances *b* and *d*, the slider of which is moved downwards by the motor spindle until $a + b$ once more equals $c + d$. The currents through C_1 and C_2 are then also equal; BR opens its contacts, the motor is stopped and the circuit is balanced once more at a new valve position. The desired value is chosen by adjusting the length of link *l*: γ_P e.g. by shifting point *f* along the slider

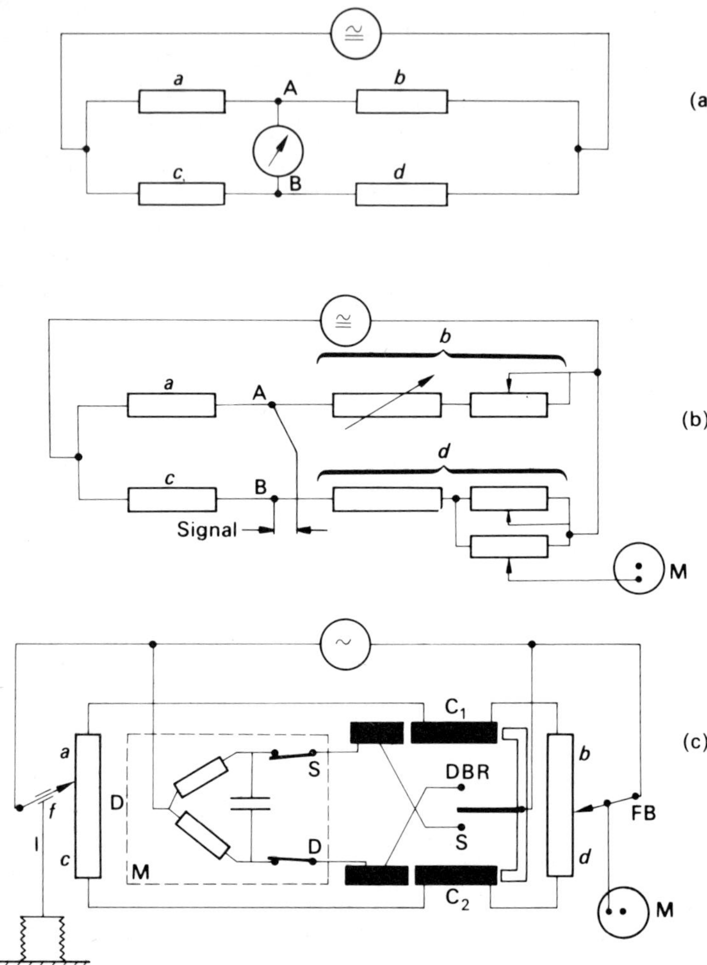

Fig. 4.10
Bridge circuits in use with electro-mechanical and electronic controllers. (*a*) wheatstone bridge; (*b*) bridge circuit such as might be used in electronic controllers; (*c*) bridge circuit in which balancing relay BR reacts to a difference in branch currents. D, sensing element; M motor.

arm of the sensing element. Many other circuits differing in minor detail from the one described but operating in a similar fashion, are in use.

Before proceeding, attention should be drawn to an interesting detail in Fig. 4.10 (*c*). When the deviation is small or, as the position corresponding to the value at which the circuit is re-balanced is approached, the difference in the forces with which C_1 and C_2 attract their respective armatures becomes very small. As a result, the relay is highly sensitive to vibrations, and the contacts tend to chatter. Therefore, the motor current is made to pass through one of two auxilliary coils C_3 and C_4. As soon as the contact is made, both the force attracting the armature and, thus, the contact pressure are effectively increased. This allows a spring to return the relay contact positively to its mid-position as soon as the attracting force is insufficient to keep the contacts closed.

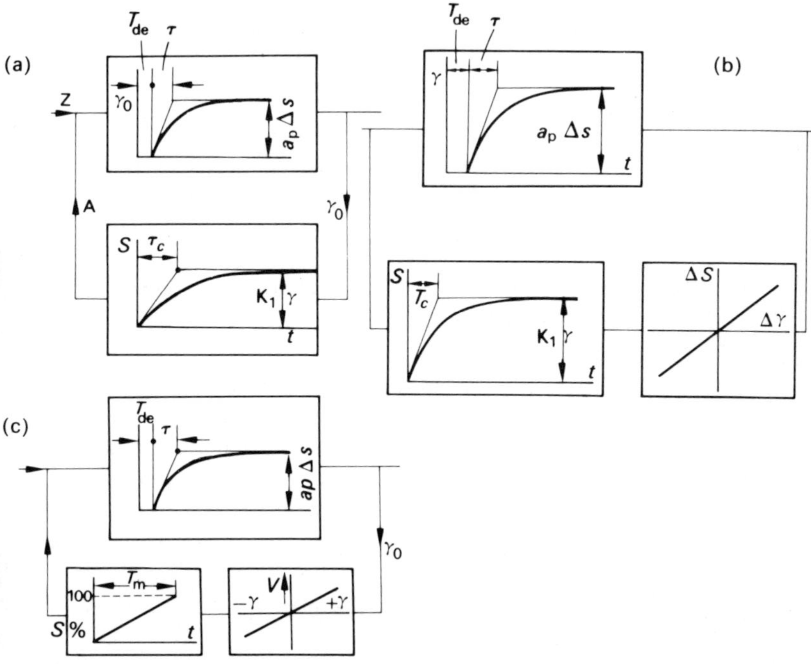

Fig. 4.11
Block diagram of a P controlled process, by means of (*a*) a self-acting controller; (*b*) a pneumatic-, and (*c*) of an electronic (or electro-mechanical) controller.

4.5. The block diagram

Figure 4.11 shown various block diagrams. The one at (*a*) refers to a self-acting controller similar to that of Fig. 4.9 (*a*). Sensing element and valve are linked so intimately that their action cannot be separated. Actually, it is the detector that is mainly responsible for the exponential response to a deviation. Part (*b*) symbolises the action of a pneumatically controlled system with an extremely fast detecting element; here it is the actuator that introduces the exponential lag τ_c. At (*c*) the block diagram is given of an equally fast electronic or electro-mechanical controller. Here, the response of the motor is linear. As shown in Fig. 4.8 the careful approach to γ_i, inherent in the exponential response, is a factor contributing to stability.

4.6. The practical application of P controllers

4.6.1. Typical applications

Reference to the opening sentence of Section 2.3 will show that the main conditions for the successful application of P controllers are similar in kind to those for two-position controllers. In a way, this is not all that surprising, as a P controller of which γ_P is reduced sufficiently acts as if it were of the two-step mode.

The range of application is very wide indeed. Especially when, providing they are constant for long periods at a time, certain departures from the desired value are permitted, stability can be obtained by increasing γ_P as necessary, and γ_i can be adjusted manually to suit ψ from time to time so as to keep the control point at a suitable value. Even in cases where γ must be kept within narrower limits, P control is used for the most diverse applications. James Watt used the principle when controlling the number of revolutions of his steam engine. Simple applications, such as the control of the cooling of motor car engines, or of water level in cisterns are frequently of the proportional mode, as are the feeding water supply systems in medium and even some large steam boilers.

4.6.2. The adjustment of the proportional band

If there is an opportunity to determine the step response of the process, and if the resulting curve permits the tangent to be drawn at the osculating point with a reasonable degree of accuracy, and if, furthermore, T_M is known, a fair idea of the stability limit can be obtained from Fig. 4.8. Alternatively, Chien, Hrones and Reswick[4] have indicated that, for optimum results, $\gamma_P = 3{\cdot}3\ T_{de}/\tau$. Frequently, however, no time is available for experimentation and near enough is considered good enough.

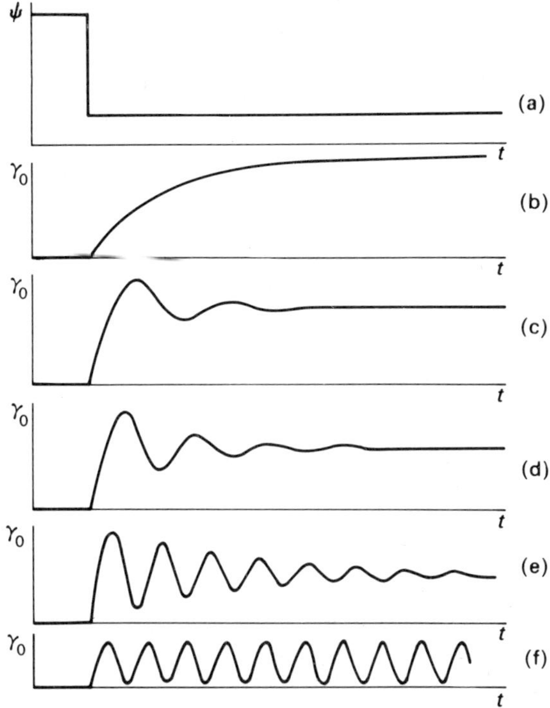

Fig. 4.12
Step response to a change in load, ψ, of a P controlled process for progressively smaller values of γ_P. At (*b*) the response is aperiodic, and the offset $\gamma - \gamma_i$ the largest; at (*f*) the system has just started to oscillate; at (*d*) the performance is very acceptable.

In these cases γ_P is reduced by steps (Fig. 4.12) until the circuit starts to oscillate as at (*f*), when a change of γ_i or other disturbance has initiated a valve movement. The γ_P at that point is then noted and the controller adjusted to between 1·5 and 2 times that value. The less favourable the conditions under which the adjustment has taken place – usually at low load – the more the margin of safety may be reduced.

4.6.3. Limit control

Limiting devices may prevent a variable, not necessarily the controlled condition, to be exceeded under any circumstances. They are usually applied for reasons of safety. Take the case of a hot water heating coil embedded in a concrete floor. If the flow temperature is exceeded the damage is so extensive that, however seldom it will be needed in normal operation, a rigid high limit must always be provided. Another high limit is frequently required with a steam heated calorifier. In both cases the limit is applied to the flow temperature, i.e. the controlled condition. This is not, or is not necessarily, the case with the rigid low limit in the air duct of an air-conditioning installation when the exhaust temperature is controlled from the space or from the exhaust duct which is mandatory in all cases where the danger of freezing up is potentially present. In these cases the authority of the limit detector is said to be 100 per cent as defined in Section 1.6.2 and elaborated below.

Flexible limitation, on the other hand, is usually applied for reasons of comfort or convenience. Thus the exhaust temperature in the last example may be limited to, say, 18°C so as to avoid complaints of a draught. Limiting devices will necessarily prevent the control system from maintaining the desired value; γ_i. Due to unusually heavy occupancy space temperature will tend to rise if, in order to keep the temperature down, the discharge temperature would have to drop to a value well below the 18°C. It may then become so hot that the occupants welcome a lower discharge temperature as a breath of fresh air which earlier, would have been objected to as a draught. In such cases it is no more than common sense to allow the low limit to be exceeded by an amount acceptable to the occupants. It is clear that the greater the rise in space temperature, the more the discharge temperature must be allowed to drop below the low limit. The ratio between the two is called the authority of the limiting device:

$$a_L = \frac{\gamma}{\gamma_{LL}} \times 100 \text{ per cent}$$

Thus, if the discharge temperature is allowed to drop 2°C for every 1°C rise in space temperature, the authority would be:

$$a = \tfrac{1}{2} \times 100 = 50 \text{ per cent}$$

The authortiy of a rigid low limit would equal

$$a = \frac{\gamma}{0} = \text{inifinity}$$

and that of a rigid high limit, *mutatis mutandis* would be the same. The authority of a flexible limiting device must be adjustable, usually on the instrument itself.

In pneumatic and many electric control schemes, the automatic limi-

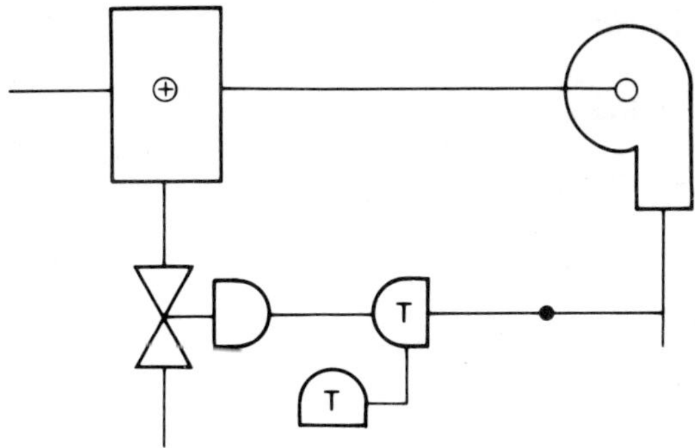

Fig. 4.13
Limitation in pneumatically controlled schemes. The signal from the detector can be passed through the flexible low-limit before energising the valve.

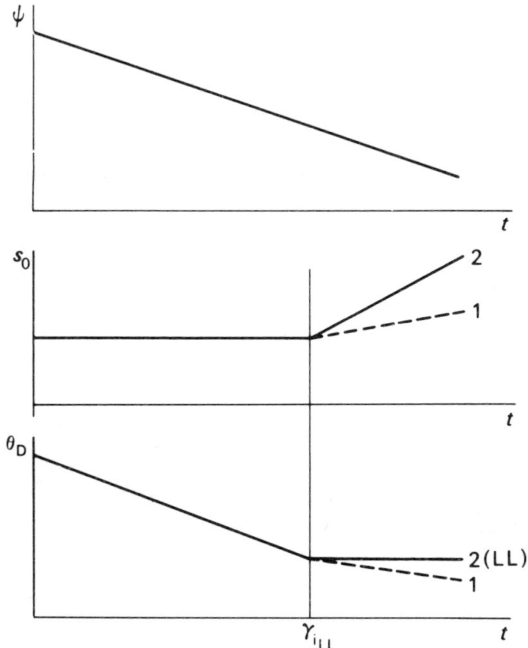

Fig. 4.14
Flexible (1) and rigid low limit control (2) and their effect on the controlled variable γ_0 and the discharge temperature θ_D in Fig.4.13.

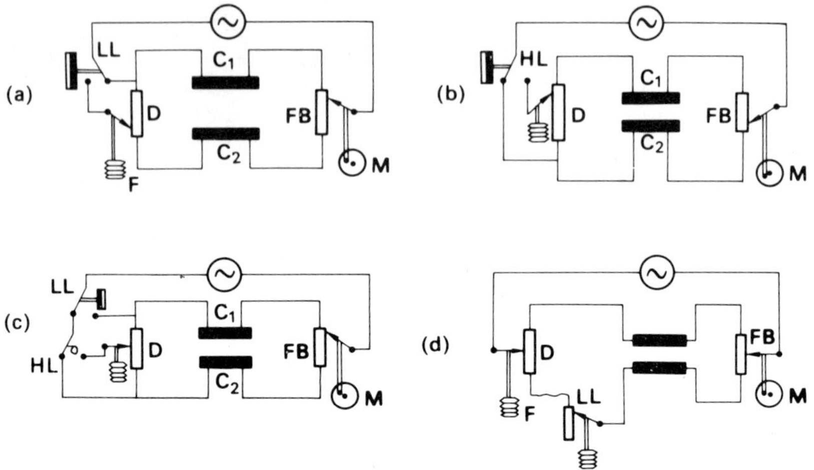

Fig. 4.15
Rigid (*a*),(*b*) and (*c*) and flexible (*d*) limitation in electro-mechanical P control.

tation provided by the master-submaster scheme, to be described in detail in Chapter 10, is in general use. An alternative method is shown in Fig. 4.13 where the signal from the space temperature detector T_1 passes through the limitstat T_2 in the discharge duct. As long as θ_0 is above the low limit, the signal is passed on to the valve actuator unchanged. However, as θ_D oversteps the low limit, the limit detector takes over and thus prevents the valve from closing further than can be tolerated under the conditions prevailing (see Fig. 4.14). In electro-mechanical systems, a fixed-limit device will frequently consist of an ordinary two-step detector. Figure 4.15 shows how it may be connected up; (*a*) and (*b*) illustrate rigid high and low limit control with traditional balancing relay schemes; (*c*) shows a low temperature and a high humidity limitation. At (*d*) a flexible low limit scheme is illustrated in which the low limit device is similar to D in Fig. 4.10(*c*), under circumstances adapted for use in air ducts.

The discontinuity introduced by all limiting arrangements is shown in Fig. 4.16 for P controllers. At (*a*) a high limit is imposed on the flow temperature, θ_F, at 70°C. If this value is not exceeded, the ambient temperature γ_0 will drop parallel to the outside temperature θ_0 (lines 4); the authority of the limitstat. as defined above, rising to infinity. As

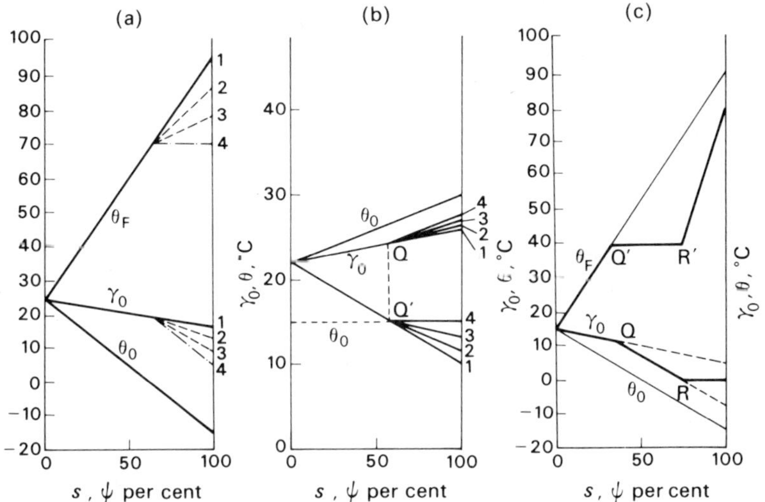

Fig. 4.16
The effect of limiting arrangements on P control characteristics (*a*) 1, 2, 3 flexible high limit control on the low temperature θ_F with increasing authority causes space temperature, γ_0, to be deflected and to run increasingly parallel to outside temperature θ_0, until, at 4, rigid limitation has been achieved; (*b*) a flexible low limit is imposed upon discharge temperature θ_D during the chilling phase of an air conditioning installation; (*c*) rigid high- and low limits on flow and space temperatures respectively in a hothouse for lettuce.

the authority of the limiting device is decreased, the original θ_F and γ_0 characteristics are approached (lines 3, 2, 1).

At (*b*), conditions during the cooling stage of an air conditioning installation are shown. Space temperature, γ_0 is allowed to rise with outside and discharge temperatures θ_D and θ_0 until 22°C when the valve opens θ_0 starts to drop, potentially to 10°C in order to keep γ_0 within a γ_P of 4°C. However, the discharge temperature is to be limited to 15°C, a value reached at Q′ when the ambient temperature, at Q, has reached just over 24°C. In the case of a rigid low limit, space temperature, γ_0 would, from that point on, rise parallel to the outside temperature. If the limit is a flexible one, γ_0 will rise according to a line between the latter and original γ_0 characteristic.

Finally, illustration (*c*) shows the combination of a rigid high limit on the flow temperature at 40°C and an equally rigid low limit at 0°C on ambient, overriding the earlier limit. Such schemes may be desirable

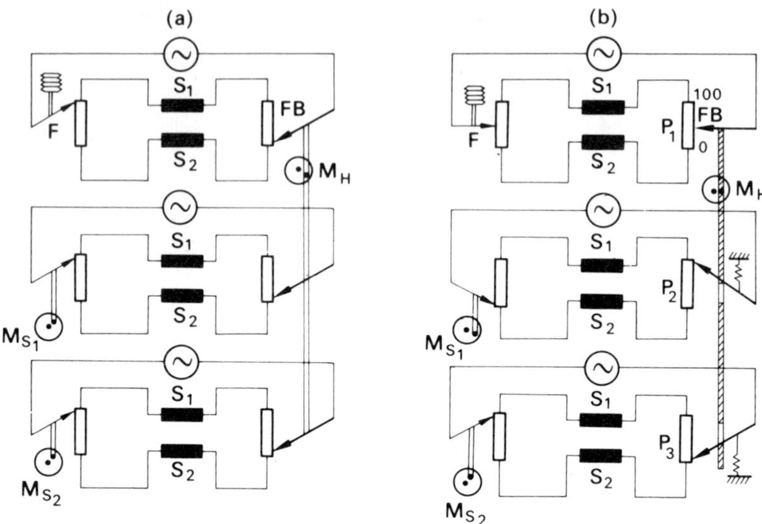

Fig. 4.17
Following and sequencing arrangements.

in hothouses in which lettuce is grown during the winter. Here the crop has to be protected both from discoloration of the foliage at high flow temperatures and from frost. At Q, after the high limit has come into operation, θ_0 starts to drop with outside temperature θ_0 until, at R′, θ_0 equals 0°C, the low limit overrides the high one, and the flow temperature is allowed to rise from point R′ on as much as is necessary to prevent frost damage.

4.6.4. Final control elements as followers or in sequence

On paper the two schemes are very similar; the difference is not so much in the components used as in their linkage. In both cases, the master motor is controlled in the usual way. Figure 4.17 shows how electrical actuators are fitted with one or two extra potentiometers, the wipers of which are rigidly connected to that of the feedback potentiometer. At (*a*), the auxiliary potentiometers operate the following motors, M_{s_1} and M_{s_2}, each via its own balancing relay. They therefore move in phase and sympathy with the master motor as far as sensitivity and hysteresis permit. Operation in sequence (*b*) also necessitates two extra potentiometers and balancing relays. The auxiliary potentiometers, however, are linked in such a way that M_{s_1} moves over the whole of its stroke and then

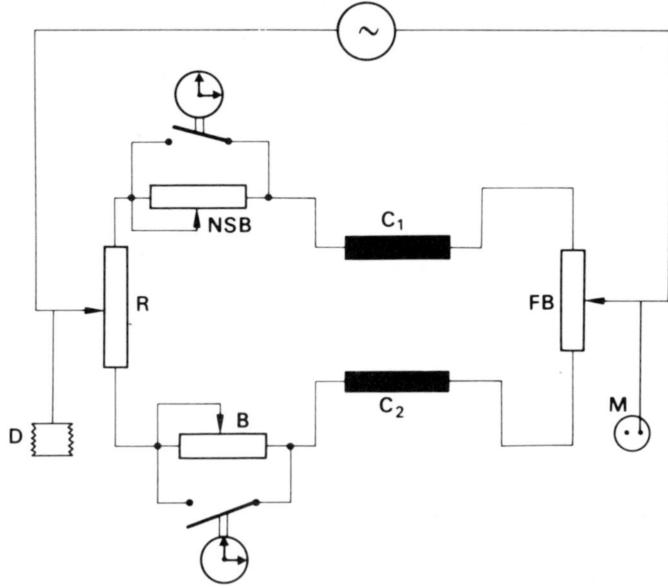

Fig. 4.18
Night set-back and early morning boost are provided by fixed or adjustable resistances, incorporated in the bridge to change the calibration. They are shorted out by the time switch when not in use.

stops when the master, starting from mid-position moves from 50 to 0 per cent. The second potentiometer p_3 is operated as the master motor moves through the other half of its stroke, i.e. from 50 to 100 per cent.

Pneumatically the scheme is more simple still. The actuators are simply supplied with the same signal. If the springs resisting the movement of the valve spindle are identically rated, the movement of the actuators will be synchronous. If sequencing is desired, the springs can be rated differently, e.g. from 0·2 to 0·5 bar (3–8 p.s.i.) and from 0·6 to 1 bar (9–14 p.s.i.). Now each actuator reacts to a corresponding range of signal values. In this case a *dead zone* corresponding to 0·1 bar, or 1 p.s.i. as the case may be, between the two steps. Provided suitable springs are available, any arrangement is possible. Thus a three-stage scheme might employ actuator springs rated 3–7, 7–11 and 11 to 15 p.s.i. or their equivalent in metric units. A very useful device in this connection, the valve positioner, is dealt with in Chapter 6.

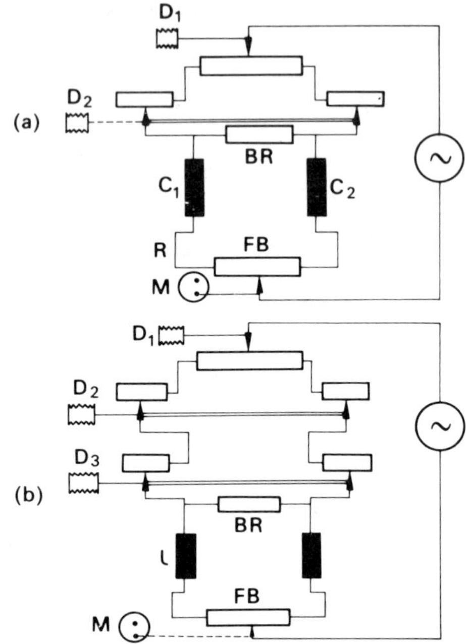

Fig. 4.19
Averaging schemes for two or three sensing elements with equal authority, as applied to an electro-mechanical P controller.

4.6.5. Day-night control

Pneumatically, this is usually done centrally by a step change in the mains pressure, e.g. from 15 to 25 p.s.i., dependent on type and origin. In electro-mechanical or electronic control schemes, resistances placed in one or other of the branches of the bridge and shorted out by a time switch contact can provide night set-back, or boost at will (Fig. 4.18).

4.6.6. Averaging control

Any second or further detector must be fitted with a double potentiometer, the wipers of which assume their mid-position when $\gamma_0 = \gamma_i$. So as to bring the bridge current back to its original value, reduced as it was by the extra resistance, a ballast resistor is placed across the combined potentiometers (Fig. 4.19). Averaging with various degrees of authority as applied to electronic controllers will be dealt with fully in the following chapter. In pneumatic control schemes a so-called averaging relay is required. This usually provides the average of two signals only which offers no difficulty as long as the number of signals to be averaged is even. If odd, for example 3, no true average is obtained with such relays unless one of the signals is reduced by a proportioning relay before being combined with the other two.

5. Proportional controllers (Electronic versions)

5.1. General

As defined earlier, all controllers in which the signal emitted by the sensing element is amplified electronically are grouped under the above heading. Their main features, in addition to their quick response and, especially, their flexibility, which will be apparent from the following paragraphs, were listed in more detail in Chapter 1, and need not be elaborated on here. Mass production methods have reduced their price relative to other types of controller and they enjoy great popularity both for the control of medium sized and larger heating and air-conditioning plants and other processes in which the controlled condition has to be maintained with high standard of accuracy; reason enough for devoting a special chapter to the features they have in common, of whatever type or make they may be.

5.1.1. Temperature sensing elements

There are two main types: those whose resistance increases with a rise in temperature, and those in which it drops with temperature rise, p.t.c. and n.t.c. resistances respectively. Both relate to the resistance value at a given temperature; in industry p.t.c. sensing element of 100 ohms at 0°C are common, but other values, e.g. 500 and 1000 ohms are also encountered. Within their range the increase in resistance is reasonably linear.

In elements with a negative temperature coefficient the change in resistance per degree is far greater. This advantage is partly cancelled out by their non-linearity, the linearising resistances R_1 and R_2 (Fig. 5.1) reducing their sensitivity. Yet, signals from n.t.c. sensing elements need considerably less amplification than those from p.t.c. wire coils.

The thermal capacity of both temperature-sensitive elements is very small. If the heated medium has free access to the wire, its larger surface-to-volume ratio gives it the advantage. When placed in a pocket its lead is largely lost and an n.t.c. soldered to a metal base plate that may be

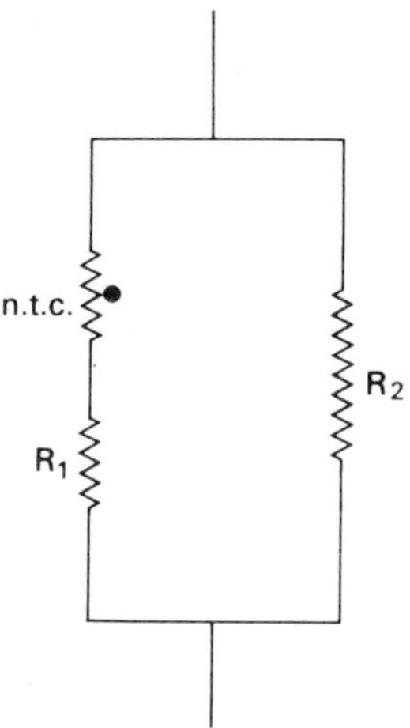

Fig. 5.1
The linearisation of n.t.c. resistances.

pressed against the bottom of the pocket is likely to respond the more rapidly. In either case the bridge current must be severely limited to prevent the resistance from changing appreciably due to the heat it generates in the temperature-sensitive element. To a certain extent this is unavoidable, a fact taken into account by calibrating them under conditions similar to those under which they will be required to operate in practice. When exposed to a fast moving airstream, such as may be encountered in some computer rooms, more heat may be carried off than normally, and the detector may tend to call for a higher ambient temperature than the value it was supposed to control at. Unless this phenomenon is understood and accepted by the client, recalibration on the spot may be necessary.

The bridge may be energised by ac or dc. As long as it is in balance the output signal is zero and the way in which it is energised is immaterial. When control action is called for, the signal is sensitive to mains fluctuations in either case, but ac bridges are the more sensitive to disturbances emanating from adjacent power lines than dc bridges, and screened cable may therefore have to be used. Limiting also is a relatively simple affair in dc bridges, as is the feedback arrangement.

5.1.2. AC or DC amplifiers

In amplifiers the designation ac or dc does not refer to the method of energising the various stages, but rather to the type of signal handled. AC amplifiers are far less sensitive to electrical interference, mains fluctuations etc., and are far more flexible than dc amplifiers. This is the reason why the shortcomings of ac energised bridges are frequently accepted; the conversion of dc signals to ac is not impossible, but it means an extra complication that many designers prefer to avoid. Frequently, also, the ac signal is rectified after amplification – for the sake of comparing it to a dc feedback signal. Sometimes the feedback potentiometer is incorporated into the bridge circuit, in which case both are energised by the same source.

5.1.3. The bridge circuit

The bridge circuits do not differ materially from those briefly discussed in earlier chapters. They also consist of a minimum of four resistances, frequently, but not necessarily, equal at calibration temperature. A number of potentiometers are usually added for purposes of adjustment.

In Fig. 5.2(*a*), the bridge is given in its simplest form. One of the resistances here marked *a* is made sensitive to the controlled variable and sized such that when $\gamma_o = \gamma_i$,

$$a : b = c : d, \text{ or}$$

$$ad = bc$$

Once more, no voltage difference is measured between A and B under these circumstances; the bridge is in balance. When a deviation develops and the resistance of *a* increases or decreases with the change in γ_o, the above relationship no longer holds good; the bridge is unbalanced and a potential difference develops between A and B. This signal is amplified and made to actuate the control valve in such a way as to counteract the effect of the disturbance that caused the deviation to develop.

The bridge as shown could be part of an I controller and is changed to the proportional mode by completing it with a feedback potentiometer, *FB*, as at *b*. The feedback signal can, of course, be applied equally well at some later stage, in which case the presence of a bridge signal need not necessarily initiate valve movement. Incorporation of a

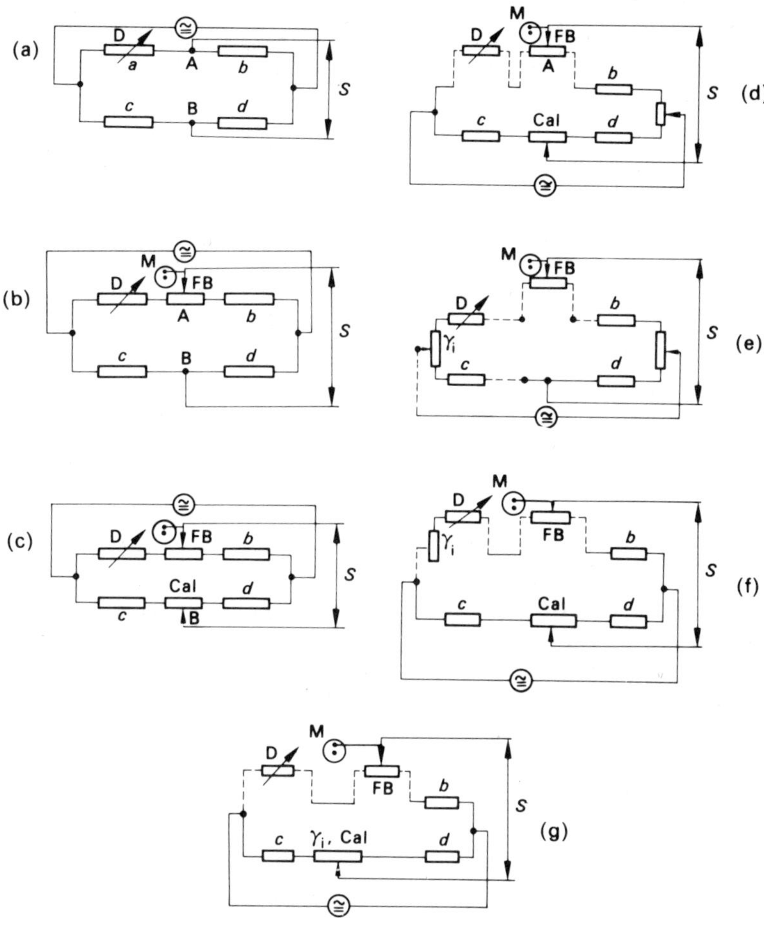

Fig. 5.2
Bridge circuits for electronic controllers: (*a*) in its simplest form; (*b*) with the feedback signal, FB, added; (*c*) incorporating the calibration in view of unavoidable variations in bridge resistance values; (*d*) illustrating the need for compensating possible errors introduced by wiring resistances, such as are avoided by the three-wire arrangement at (*e*); (*f*) local adjustment of γ_i introduces an erratic resistance into one of the branches and should be avoided; (*g*) all three variables adjusted or compensated, by means of one potentiometer. D, sensing element; A,B, points between which the signal S appears; *a, b, c, d,* bridge resistances; γ_i, Cal., adjustment potentiometers; M, motor; S, output signal of the bridge.

second potentiometer, γ_i, accessible to the user, allows a standard sensing element to be used for a range of γ_i values (*c*).

In circuits of this type, the signal, being but a small fraction of the bridge voltage which is usually of the order of 12 or 24 V, is generated by relatively small changes in the resistance values of one or more branches of the bridge. Therefore care should be taken to eliminate or compensate for any random resistances, such as those introduced by the wiring to the detector, to the feedback potentiometer, etc. that might noticeably affect the bridge currents and, therefore, the calibration. The variable resistance between wiper and winding of the FB and γ_i potentiometers can be made to influence both bridge arms equally by shifting points *A* and *B* to a junction (*c*). In arrangements to Fig. 5.2(*d*), however, the sensing element wiring affects one arm only. It therefore has to be compensated for by a further potentiometer, to be adjusted when commissioning, but otherwise not normally accessible to the user (*d*) As an alternative, resistance *c* is incorporated in the sensing element (*e*), as well as potentiometer γ_i', required in any case to iron out small variations in resistance of D and *c* in the factory. If this is made accessible to the user and fitted with a calibrated scale for local adjustment, it can be used as an alternative to the central adjustment γ_i, in the controller, which then has to be set in mid-position. This facility is lost to two-wire sensing elements as otherwise the variable resistance of the wiper-coil constant would be moved from the supply circuit to one of the bridge arms (Fig. 5.2(*f*)).

If, for practical purposes, local adjustment of γ_i may not be feasible unless the sensing elements is of the three-wire type, provision of a desired value adjustment in the controller at least offers protection against unauthorised interference; it is also useful when the sensing element is placed in an inaccessible position and facilitates averaging control. At which bridge junction the adjustment potentiometer is located is immaterial; its function, to affect the balance point by increasing one branch resistance at the expense of the adjacent one, can be achieved at any one of the four corners. Thus, all three functions of the adjustment, namely compensation for incidental wiring resistances, for minor variations in the bridge components and the adjustment of the desired value can be carried out by one single potentiometer (Fig. 5.2(*g*)). The only disadvantage of this otherwise attractive method is that each controller has to be visited by an expert for calibration purposes when it is commissioned. This means in practice adjustment of the γ_i scale to its correct position.

5.2. Adjustment of the proportional band

The comparison of the signal relating to the deviation and to valve position has been seen to be an essential part of any proportional controller.

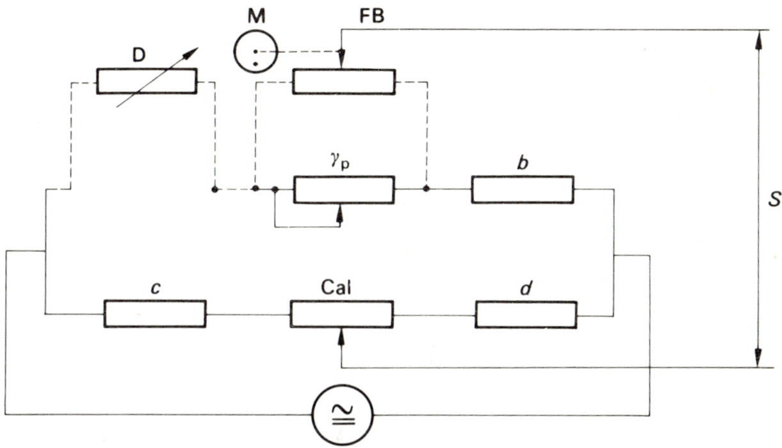

Fig. 5.3
The adjustment of the proportional band, by means of a variable resistance, γ_P, shunting the feedback potentiometer FB.

Figure 5.3 shows a common method of varying the size of the feedback signal as it influences the bridge output. The feedback potentiometer FB is shunted by a variable resistance γ_P thus reducing its effective resistance R_r to a value

$$R_r = \frac{FB \times \gamma_P}{FB + \gamma_P}$$

in which the magnitudes of these symbols are given as resistance values. This reduces the resistance values added to the two adjacent branches of the bridge in the same ratio as well as the influence on the balance point of a change in valve position. Alternative methods of adjusting the proportional band leave the feedback signal unchanged, but vary the bridge output when offered to a subsequent stage of the amplifier.

5.3. Discrimination and amplification

Though the application of contactless switching methods is increasing, the majority of amplifiers in present use are terminated by one or two output relays providing an s.p.d.t.contact, the direction of the valve movement being co-related with the position of the contact. The function of the discriminator then is to affect the signal S in such a way that the position of the relay contact corresponds to the sign of the deviation.

This can be done by utilising either the sign, the magnitude, or the phase of the signal or a combination of them. Since the introduction of transistor amplifiers and, more especially of integrated circuits, the number of possibilities has increased to such an extent that to describe even a small proportion of them would read more like a treatise on electronics than on control. For this reason both discriminator and amplifier will be assumed to be given entities: at most, such details are of interest when defective controls have to be repaired and do not affect the results.

5.4. Practical application of electronic P controllers

5.4.1. Combining two or more signals

Any arrangement of sensing element used for averaging must ensure that the bridge equation

$$ad = bc$$

is maintained. Thus, if, as in the simplified bridge circuit of Fig. 5.4 (*a*) two detectors are placed in series, their resistance must be one half, in parallel (*b*) it must be double their usual value. For arrangements as at (*c*),(*d*) and (*e*), standard sensing elements may be used.

It should be realised that locating various sensing elements in such a way as to take into account random variations in the controlled variable, and preventing any one of them from obtaining unwarranted importance may improve conditions to a certain extent, but do not cure the basic fact that the process is not homogeneous. No corrective action can meet the requirements of each part. It is better to try to improve the homogeneity of the process than to rely on methods like averaging that, though in some cases indispensable, are nothing but palliatives when all is said and done.

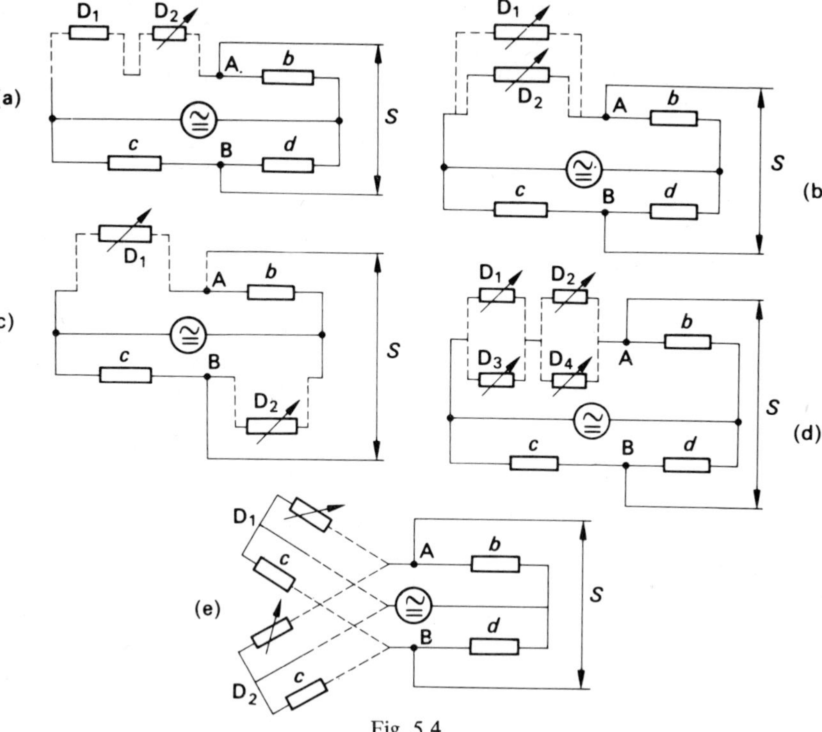

Fig. 5.4
Averaging control. Two wire sensing elements of suitable resistance may be placed in series (*a*); in parallel (*b*), in opposite branches (*c*) or as a combination of (*a*) and (*b*). Averaging with three-wire sensing elements is shown at (*e*).

5.4.1.1. Introduction of factors other than γ_0 may improve control quality. The offset depending, as it does, on the load, ψ, can be reduced, removed or overcompensated by adding a signal proportional to ψ to the input. In this way the shortcoming of the proportional system limiting the corrective action to a value short of requirements, and thus causing an offset even though the capacity to remove it is available, could theoretically be eliminated. In practice, this method is seldom completely successful as it is difficult to measure the load accurately enough in one location; the factors of which ψ consists enter the process and influence it at various stages. Therefore, such an open-loop component cannot replace good control by closed-loop methods but can certainly improve it. This is most effectively done by reducing the span to a value commensurate to ψ. Itself unable to cause instability, such an open-loop component allows the proportional band to be reduced without endangering stable operation.

Applied to heating and air-conditioning plants, open-loop elements are also used to raise the control point with load during the winter for reasons of comfort or for the sake of economy during the summer period. In the latter case, the natural offset is actually increased on purpose rather than reduced.

In these as well as in several other cases, to be discussed further in the following sections, means must be found to relate the influence of such and other signals to that of the detector. This relationship is expressed by the concept of authority, which now needs definition.

5.4.1.2. The authority of sensing elements and of their signals. When a sensing element has established a deviation equal to the proportional band, γ_P, and has caused the final control element to run through the whole of its stroke, $\Delta s = s$, the authority a of that particular sensing element is said to be 100 per cent. Figure 5.5 which is best understood by assuming γ_i to correspond to one of the extreme values of γ_P, at which the valve also assumes one of its extreme positions, illustrates this concept by means of a balance. The lengths of the balance arms are made to represent the authority, a, and γ_P. The two forces, are made to correspond to the deviation γ and the valve movement expressed as a fraction, Δs of the stroke. In this case (Fig. 5.5 (*a*)) unity, or 100 per cent, the value will travel through the whole of its stroke. Should a be reduced to 50 per cent, γ must be doubled to obtain the same valve movement (Fig. 5.5 (*b*)) or, if both γ and a remain constant (*c*), the valve movement will be halved. In general

$$\frac{\Delta s}{s} = \frac{a\gamma}{\gamma_P}, \text{ per cent}$$

If the deviations as measured by two sensing elements (*e*) and (*f*) are of the same or opposite sign, the products with their respective a-values are added or subtracted as the case may be. If the left-hand balance arm is weighted by the signals of three equal sensing elements of equal authority, that of each will contribute as if it alone were operative (Fig. 5.5(*g*)).

In Fig. 5.5 (*h*), two sensing elements with authority values of 100 and 10 per cent respectively are working in sympathy with one another. If the first measures a departure of 1°C and the second of 5°C, the

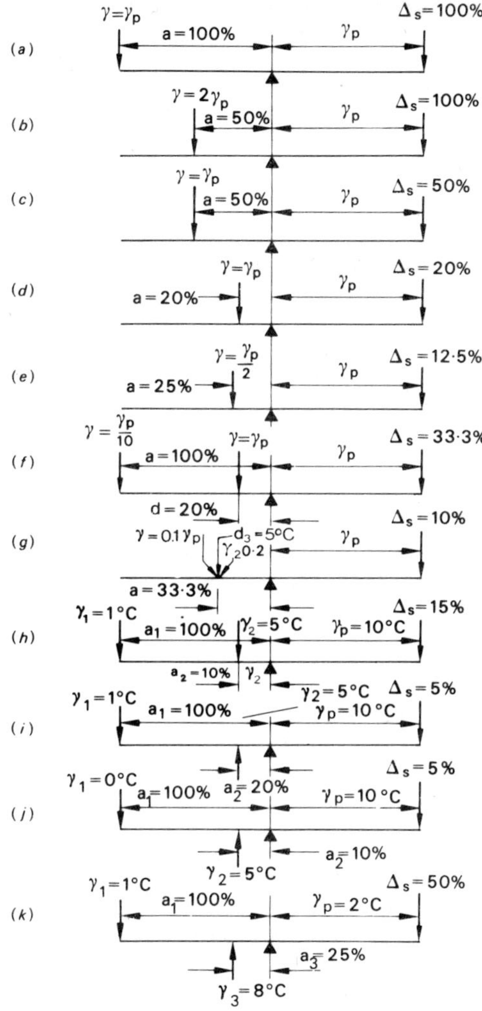

Fig. 5.5
The authority of one or more sensing elements, illustrated by the analogy of a balance.

resultant valve movement at a $\gamma_P = 10°C$ will equal

$$\frac{\Delta s}{s} = 0{\cdot}1 \times 100\% + \frac{5}{10} \times 10\% = 15\%$$

The signals of the two detectors could have been made to oppose one another (*i*), in which case

$$\frac{\Delta s}{s} = 0{\cdot}1 \times 100\% - \frac{5}{10} \times 10\% = 5\%$$

In all these cases the signals derive from closed-loop systems. Should one of the detectors (D3) be placed outside, an open-loop element is introduced which causes a shift in desired value. Let the control constants of a heating installation in a cold climate ($Sp = 40°C$) allow stable operation at a γ_P of 5°C. It is desired to reduce the offset to zero, in as far as this is due to outside influences. At 20°C outside temperature no action is taken; at −20°C the desired value must be raised by the value of γ_P so as to allow the valve to be fully open notwithstanding $\gamma_1 = 0$. The authority of D_3 must therefore equal 5/40 or 12·5% (*j*). If a cooling installation is capable of maintaining an ambient temperature of 24°C at an outside temperature of 30°C, but it is wished to reduce the difference of the two to 4°C ($\gamma_i \rightarrow 26°C$), at a nominal $\gamma_i = 22°C$ at zero load and a $\gamma_P = 2°C$, a change in outside temperature of 8°C (i.e. 30 − 22°C) must shift the inside temperature by $26 - 22 - \gamma_P = 2°C$ and the authority of D3 must be adjusted to

$$a_{D3} = \gamma_{D1}/\gamma_{D3} = \tfrac{2}{8}\,100\% = 25\%$$

This case is illustrated in Fig. 5.5 (*k*).

5.4.1.3. The authority adjustment There are various methods of adjusting the authority. The arrangement most widely used is that shown in Fig. 5.6 (*a*) in which each detector is assigned its own bridge circuit; only part of the output of the one with reduced authority is added to the output of the main bridge, and a potentiometer placed across the second bridge allows *a* to be decreased to any value between 100 and 0 per cent. At (*b*) the two bridges have the branches containing the fixed resistances *c* and *d* in common, and the relationship of the authority values is given by the relative parts of the output signals applied to the amplifier as input. Varying the bridge voltage as at (*c*) is also a convenient method of achieving similar results. The method by which authority is reduced by fixed series or parallel resistances (*d*, *e*) though simple, lacks flexibility unless the resistances are made adjustable.

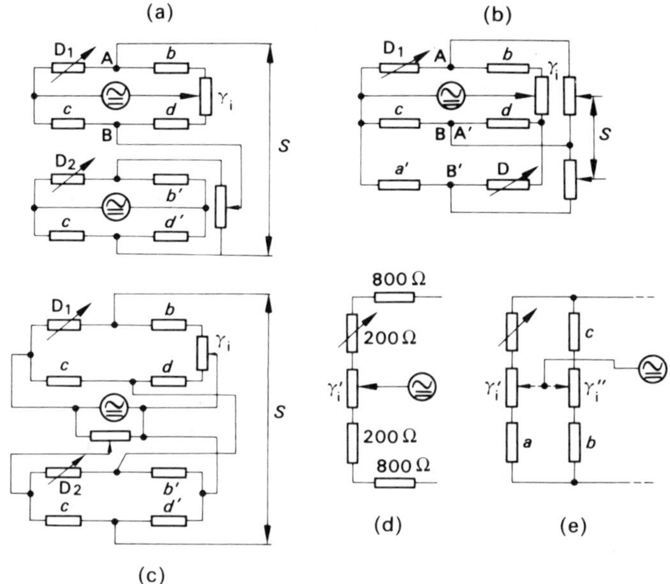

Fig. 5.6
The authority adjustment of a detector.

5.4.2. Limit control

Owing to the rapid response of 'electronic' detectors, rigid limiting devices are not often called upon to come into action. This should by no means be taken to mean that they are never needed. They usually take the form of ordinary two-step sensing elements as was the case with electro-mechanical and pneumatic P′control circuits, frequently wired to a multi-pole relay. Use of limiting devices as described in Chapter 3 reduces the fluctuations introduced by two-step detectors (Fig. 5.7 (*a*) and (*b*)). For flexible limit control also, electro-mechanical devices are favoured (Fig. 5.7 (*c*)), although electronic sensing elements are also in use. In the former case it is adjusted in such a way that no resistance is introduced into the circuit until γ_0 starts to transcend the limit in the direction considered undesirable. As its standard resistance (135 ohm) is far in excess of the resistance change in an 'electronic' sensing element it must be shorted by a low-ohmic potentiometer which can also be used for the adjustment of the authority.

Electronic sensing elements lack this natural advantage of one-way action when used as a limit. If, as low limit, their signal opposes that of the main detector below a given value, it will increase the signal from

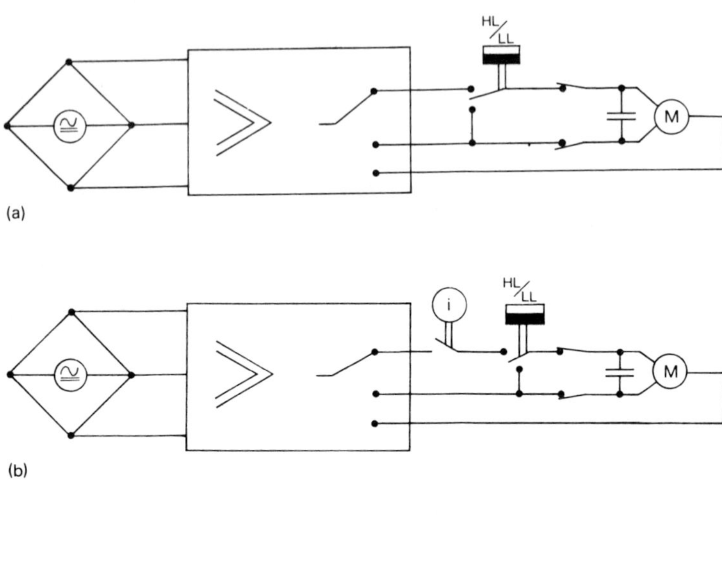

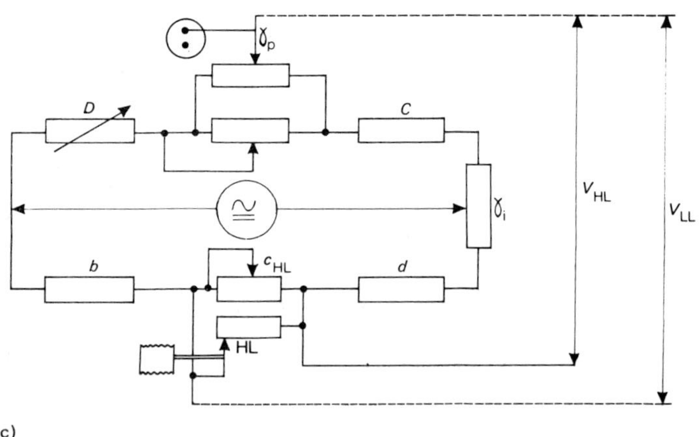

Fig. 5.7
Rigid, (*a*) and (*b*) and flexible limitations (*c*) by means of electro-mechanical devices. In the case of a flexible limit by means of a P sensing element the authority is adjusted by means of a shunt resistor, a_{HL}. HL, high limit detector.

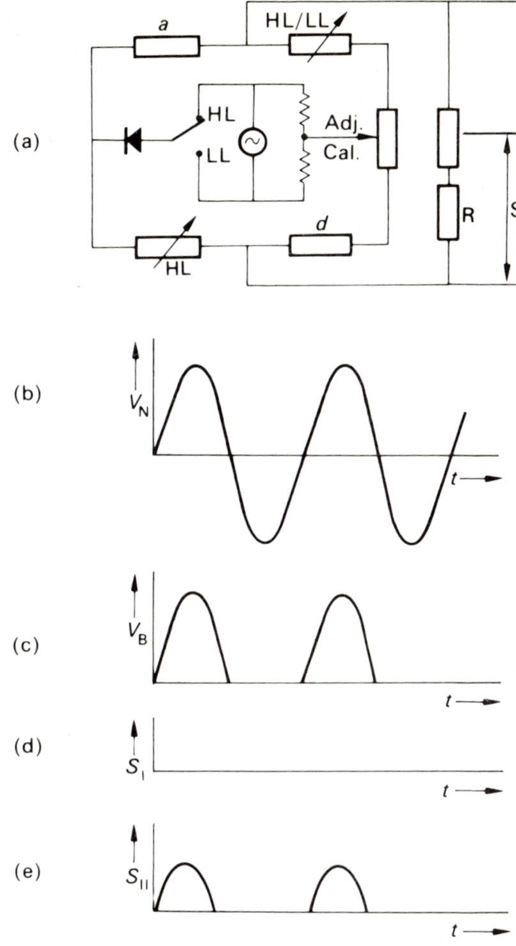

Fig. 5.8
Flexible high or low limit control by 'electronic' sensing elements. The double rectification incorporated in the bridge (*a*) reduces the a.c. bridge voltage (*b*) to a series of impulses (*c*) which can generate no signal as long as γ_0 is on the safe side of the limit (*d*), but makes itself felt (*e*) as soon as the limit is transcended.

the controlling sensing element above that value and thus cause the control point to fall unduly. The bridge is energised by a.c. so that the desired action may be obtained by the use of rectifiers (Fig. 5.8). Electronic limiting devices are usually, but not necessarily, applied in a

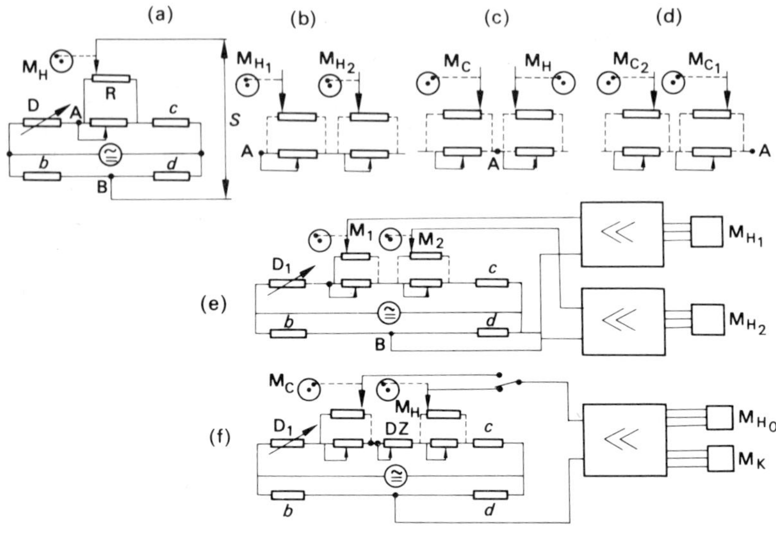

Fig. 5.9
Sequencing of two final control elements (*a*) only one valve is present; (*b*) hooking up the feedback potentiometers of two heating (*c*) one heating and one cooling and (*d*) two cooling stages; (*e*) the signals from the two stages can be amplified separatly or together (*f*). In the latter scheme an adjustable dead zone is included.

separate bridge, the output signal, *S*, of which is placed in series with that of the main bridge. In Fig. 5.8 the auxiliary bridge can be changed from a high to a low limit arrangement by reversing the phase of the supply voltage; the signal is increased by fitting two temperature – sensitive windings connected in opposite branches of the bridge. Only when their signal is in phase with the rectified bridge voltage can the limit oppose the main signal. Resistance *R* is placed in series with the authority potentiometer so as to prevent the limit being put out of action by accidentally reducing its authority to zero.

5.4.3. Sequencing and following schemes

As was the case when this subject was dealt with in Chapter 4, the feedback schemes needs some consideration. Figure 5.9 (*a*) shows a simplified bridge circuit for a single heating valve no different from several discussed earlier. When, in balance no voltage is measured between the wiper and point A on the one hand, and point B on the other: the 'balance' point, i.e. the point on the upper branches of the bridge at which the voltage equals that at point B corresponds to point A. As, however, the sensing element temperature, and therewith its resistance, drops, the balance point wanders to the right and a signal appears between A and B. This causes the valve to be opened and the wiper to follow the balance point. Not until, within the limits of sensitivity and hysteresis, this balance point is reached, is the bridge rebalanced at which point the output signal, *S*, disappears. At (*b*), (*e*) and (*d*) three possible arrangements of the feedback potentiometers are shown. At (*a*) two heating valves are sequenced. As the balance point wanders along the two potentiometers from left to right, it is followed first by the left-hand, then by the right-hand wiper until the second valve is also fully open.

These multi-stage plants offer an opportunity of pointing out an interesting fact that, so far, has gone unnoticed. As in Fig. 4.10, the points A and B between which the signal is measured in Fig. 5.4 and 5.6 were fixed. When in the two latter cases the bridge is at balance the signal does not disappear, but is counteracted at a later stage of the amplifier. In Fig. 5.3, point A is located on the wiper; in other words, point A moves along the feedback potentiometer in sympathy with the slider movement. It does not come to rest until the bridge is balanced and the voltage difference between A and B is effectively reduced to zero.

In a heating and a chilling stage (Fig. 5.9 (*c*)), point A corresponds to the balance point when, γ_0 equalling γ_i, both motors, M_H and M_C have closed their valves. Their respective potentiometers then assume extreme but opposite positions. As the demand for heat increases, the M_H wiper will follow the balance point to the right; when chilling is required the M_C wiper will follow point A to the left. A two-stage chilling arrangement (*d*) is similar to a double-stage heating installation (*b*) but for the position of the wipers when the valves are closed. As first one, then the other valve opens up, the balance point will wander along the upper branch of the bridge from one extreme to the other. In these schemes each stage can be given its own amplifier (*e*) or the two can be combined (*f*), the motors being linked as shown in Fig. 4.17, the signal being transferred as either M_H or M_C is in operation.

An adjustable dead zone can be incoporated by placing an adjustable resistance, *DZ* between the two *FB* potentiometers. As point *A* moves along *DZ* no control action results. Such a dead zone is added when it is desired to prevent the chiller from being taken into operation until the ambient temperature in summer, i.e. in the cooling stage, has risen a few degrees above that maintained in the heating season.

In follower control the slave motor is linked to the master by means of a balancing relay, as described in Section 4.6.3.

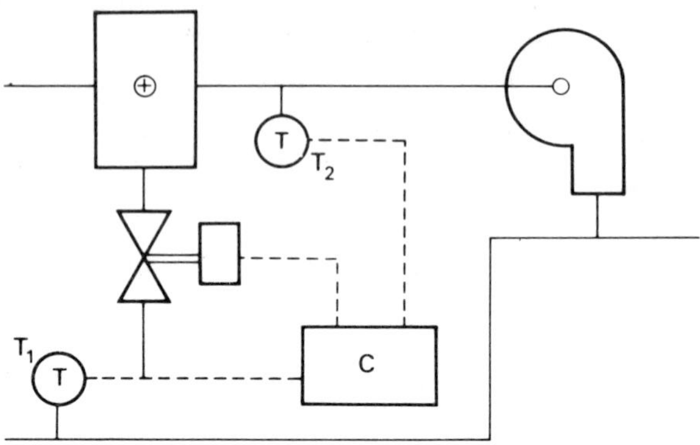

Fig. 5.10
The dual thermostat in a plenum ventilation installation controlled from the exhaust duct.

5.5. Dual thermostat

It was shown in Chapter 4 how the p.b. cannot be reduced at will without the control scheme becoming unstable. This means that the user may have to accept an offset larger than his requirements would permit. Several schemes have been evolved by which the apparently rigid link existing between γ_P and the control constants can be severed, or at least modified. One of these was dealt with in Section 5.4.1.2; a different approach is followed in the scheme to be dealt with here.

It was explained that instability occurs when the corrective action taken by the controller consistently causes a deviation greater than the deviation it was designed to reduce. The phenomenon, enhanced as the control range increases, is intimately linked to the presence of pure or effective dead time, as, during its elapse, the controller cannot intervene to adapt supply to the demand obtaining at any given moment. The only way to prevent excessive overshoot that has been discussed so far is to reduce effective span, or to reduce the movement of the final control element for a given departure. This is synonymous with widening the differential and increasing the potential offset which, at large values of T_{de}, may become unacceptable. An open loop element is only effective with respect to influence to which it is exposed.

The introduction of a so-called 'snubber' thermostat, T_2, placed, say, immediately after the battery of Fig. 5.10 reduces the movement of the final control element without increasing the offset unduly in cases

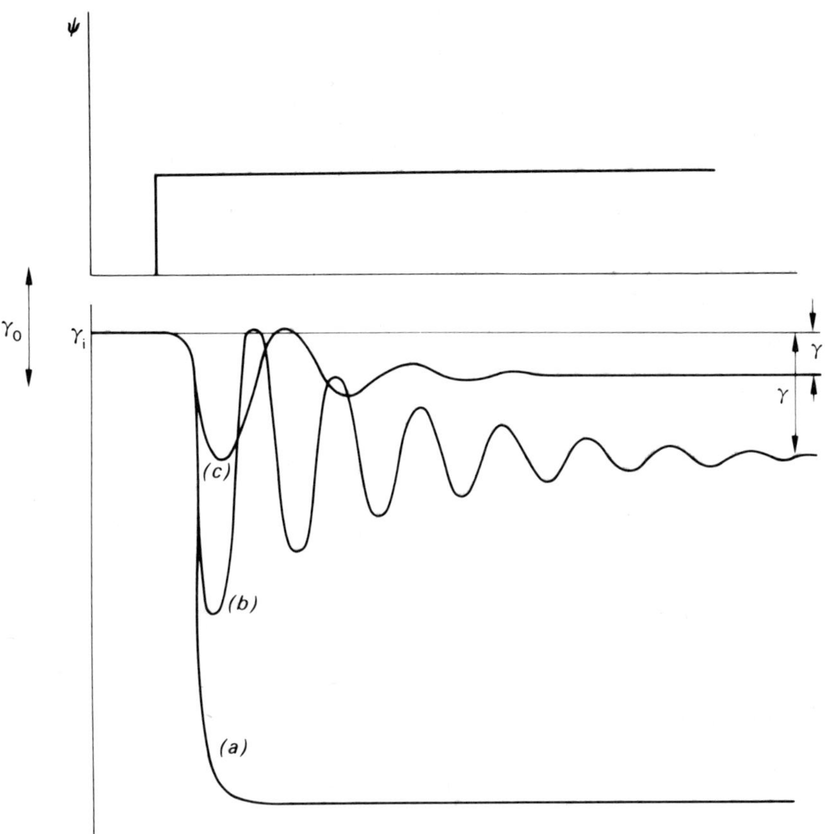

Fig. 5.11
The influence of a snubber (*a*) reaction of γ_0 to a change in load, ψ, in the absence of a controller; (*b*) reaction of γ_0 controlled proportionally at a near critical value of γ_P; (*c*) the reaction of a two-thermostat team adjusted to a value of γ_P that is smaller still.

where conditions are too unfavourable for control by a single detector. T_2 is linked up in such a way that its signal opposes any departure from its desired value usually coincident with the lowest discharge temperature permissible.

An example will clarify the way the system works. Let T_1 detect a departure which a change of valve position of 5 per cent would eventually correct or reduce. The gain, however may have been adjusted to such a value that Δs is five times that amount, i.e. 25 per cent. Due to the various lags, the sensing element cannot prevent considerable overshoot. *The snubber*, however, opposes the action of T_1 and restricts the valve movement to, say, 10 per cent. Correspondingly more time is needed to return γ_0 to its ultimate value, but the overshoot is far less than it would have been in the absence of T_2. As T_1 detects the fact that γ is diminishing and reduces the valve displacement accordingly, the snubber relaxes its opposition and the valve ultimately comes to rest not far from the position it would have assumed had T_1 alone been present (Fig. 5.11).

The action of the snubber therefore is to cause an apparent increase in proportional band restricting large departures in valve position from the one corresponding to its desired value, but to relinquish most of its opposing influence as that position is approached. It will be clear that the action of T_1 must remain the dominant factor, in other words that the authority of T_2 must be a fraction of that of T_1. Also, it must be realised that T_1 is the sole initiator of control action as far as the controlled condition is concerned; it can be opposed, but never eliminated by T_2. In so far as T_2 initiates control action, it tends to keep the discharge temperature at a constant value. Thus it may counteract the effect of a change in flow temperature or of that of the air entering the heater, in which its action is similar to that of a submaster in circuits, to be discussed in Chapter 10; this, the dual thermostat is emphatically not, neither is it a limiting device or a frost-protecting thermostat, as its position in the duct might lead one to expect.

Returning to the value of γ_P of the two thermostat team, this is increased by the action of T_2. On the other hand the stability has been improved to such an extent that the differential of T_1 can be reduced to a fraction of its original value. Thus the net effect of the snubber action is a noticeable reduction in effective proportional band, presupposing of course, that the authority of T_2 is chosen to best advantage. This is done as follows: As mentioned, the snubber is adjusted to the lowest permissible discharge temperature, say to 18°C. The heater battery is considered to consist of two separate parts. The least important one is sized to bring the air, entering the heater at its lowest temperature, up to the snubber's desired value. The second part of the battery brings the air temperature from the 18°C to its maximum value. Below 18°C the two thermostats co-operate in opening up the valve through which the battery is supplied. above its desired value T_2 will start to oppose the signals from T_1. The larger the span of the second half of the battery, the smaller T_2's authority should be chosen. Thus, if $Sp < 15°C$, a_{T2}

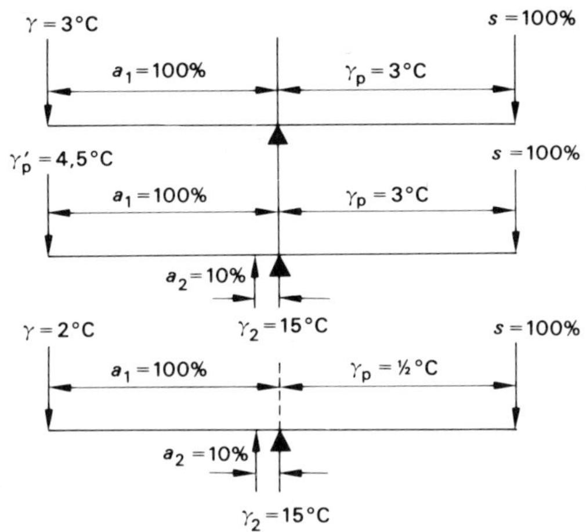

Fig. 5.12
The determination of the effective p.b. of a two-thermostat team.

can equal e.g. 10 per cent; if $15°C < Sp < 25°C$, a_{T2} should be reduced to 7·5 per cent and if $Sp > 25°C$, a_{T2} should not exceed 5 per cent. The values for Sp have to be measured at maximum load, i.e. at design temperature and fully opened valve.

5.5.1. The resulting proportional band

This is determined by the method described in Section 5.4.2. Assume that, without T_2, stability can be attained at $\gamma_P = 3°C$ (Fig. 5.12 (*a*)). If the span of T_2's heating stage is taken to be 15°C, a_{T2} can equal 10 per cent, which will, as (*b*) shows, increase the effective proportional band γ_P' to 4·5°C:

$$\gamma_P' = \gamma_P + (Sp \times a_2) = 3 + 15 \times 0{\cdot}1 = 4{\cdot}5°C.$$

The stabilising effect of the snubber, however, allows the throttling range of T_1 by itself to be reduced to 0·5°C which reduces the effective throttling range to 0·5 + 1·5 = 2°C; a reduction of 33 per cent. This arbitrary example does not necessarily show the optimum results obtainable by the introduction of a snubber.

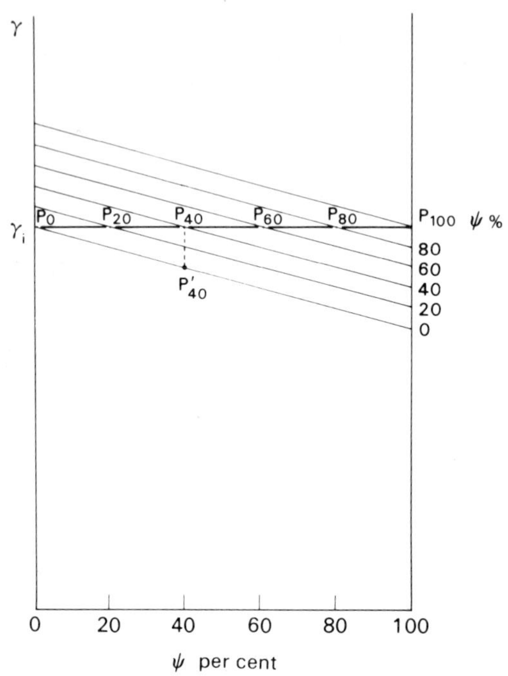

Fig. 5.13
Addition of an open-loop element could theoretically eliminate the offset: as the load ψ increases so the desired value moves from P_0 through P_{20}, etc. to P_{100}. P'_{40} is the control point at $\psi = 40$ per cent in the absence of T_3.

5.6. Further reduction of the offset

The next step is obvious: The method of the previous section can be combined with that described in Section 5.4.1.2. From a measurement of the load a signal could be derived which, suitably applied, could be made to raise the desired value by exactly the amount of the offset.

This could work as shown in Fig. 5.13. It will be observed that the figure differs from Fig. 4.3 and others, inasmuch as the abscissa has been changed from valve position, s, to load, ψ. As the load increases the control characteristic is shifted upwards by an amount equal to the respective offset, so that the desired value, in moving to the right, remains constant. The effect is strongly reminiscent of a P + I controller, to be dealt with in Chapter 8, but differs from it in certain important aspects.

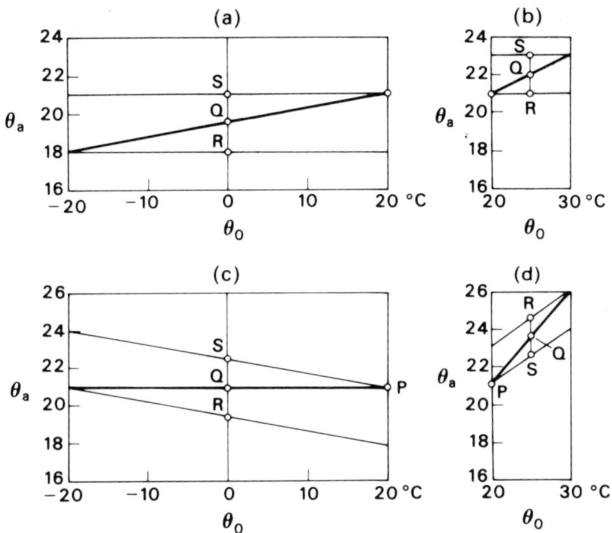

Fig. 5.14
The reduction of the offset due to external influences during the winter period (*a*) and (*c*) and the raising of the control point during the summer (*b*) and (*d*), both by means of an outside sensing element, T_3. Internal influences can still move the control point throughout the space between the thin lines upon which R and S are situated.

At first sight this is a very attractive scheme. Unfortunately, it needs qualification. If the near perfect result of Fig. 5.13 were a real possibility, why should one go to the trouble of modifying a closed-loop system by an open-loop element when a pure open-loop scheme which, as will be shown in Chapter 9, is the simplest of all control methods, would do as well? The answer is, of course, that more often than not the basic assumption, that the load can conveniently be measured, cannot be realised. Once more taking a heating plant as an example, one would be inclined to take the outside temperature as a measure of the load. In practice this is frequently done: a detector, T_3, placed on an outside wall or in a fresh air duct will emit a signal related to ψ. Actually, T_3 only measures part of the load with any degree of accuracy, namely the outside air temperature. The effect of solar radiation and ventilation losses due to wind are largely or wholly neglected. Yet these two factors can, dependent upon window size and building construction, easily cause the load to fluctuate by 40 per cent or more. Another important factor not taken into account is the internal load due to occupancy,

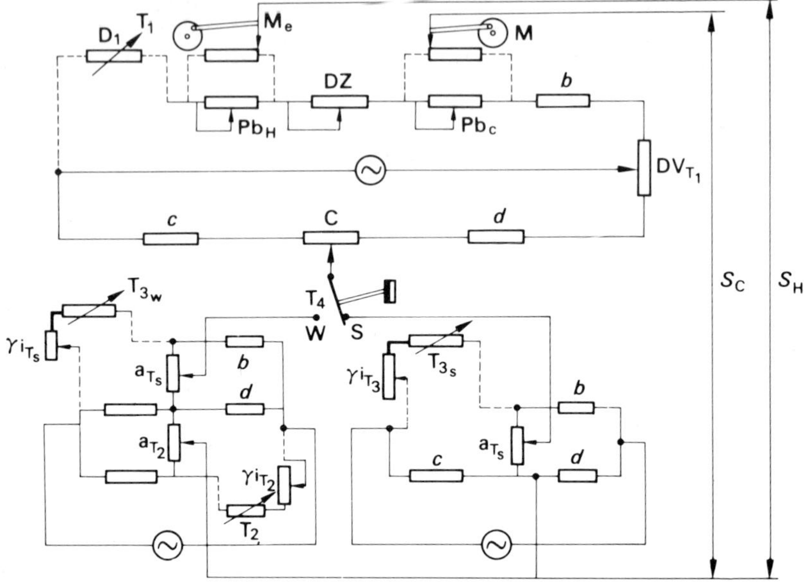

Fig. 5.15
Arrangement of bridges for a triple-thermostat with summer and winter compensation and dead zone DZ; a_{T_2}, a_{T_3} authority adjustments; γ_{iT_1}, γ_{iT_2}, γ_{iT_3} desired value adjustment; (W), (S), winter and summer positions of T_4 etc.

lighting and other sources of heat. This cannot be corrected by a further detector unless it were capable of interpreting the offset as deviation – and this would turn the control action into a P + I controller.

If, therefore, in a given plant, outside temperature accounts for, say, 50 per cent of the load, T_3 can at best compensate 50 per cent of the offset. Though not ideal, its addition is certainly worth while and the method is widely used. The authority of T_3 depends upon the amount of the offset to be compensated. Let the design temperature of a heating installation be $-10°C$, and the desired space temperature be $+20°C$, and the controller be a adjusted to a γ_P of 3°C. If the outside temperature accounts for 50 per cent of the load, the offset due to outside temperature is 3°C × 50 per cent, and a_{T3} follows from the fact that, when exposed to the full range of outside temperatures, T_3 must initiate a signal equal to the part of the offset it is supposed to compensate. Therefore, T_3 must initiate a signal in this case

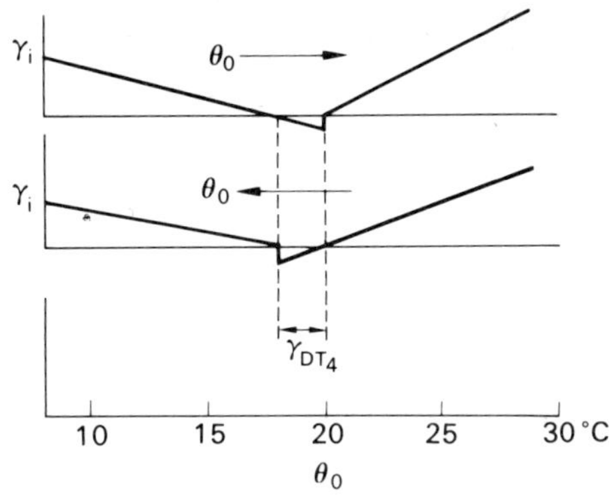

Fig. 5.16
The difference in the way the desired value γ_i changes with rising (*a*) and falling (*b*) outside temperature θ_0 due to the differential of T_4

$$\{20-(-10)\}^\circ C \times a_{T3} = \tfrac{3}{2}{}^\circ C$$

or

$$a_{T3} = \tfrac{3}{2} \times \tfrac{1}{30} \times 100 \text{ per cent} = 5 \text{ per cent}$$

5.6.1. The triple thermostat

This frequently encountered scheme is nothing but a dual thermostat to which a T_3 has been added, partially to compensate the offset as described (Fig. 5.14).

5.6.2. Summer and winter compensation

In areas with really low winter temperatures, the window panes can become so cold that the radiation temperature θ_r in the controlled space may drop to an uncomfortable level. Space sensing elements react mainly to ambient temperature θ_a. If θ_a is kept constant, the effective temperature, θ_e, within limits equal to

$$\theta_e = \frac{\theta_r + \theta_a}{2}$$

will drop with θ_r, and the occupants will complain of the cold, the more so the nearer they are to the windows.

Provided θ_r and θ_a do not differ too much, θ_e can be raised to the desired value by increasing θ_a, in other words, by raising the desired value in direct proportion to the drop in outside temperature, θ_0. The only difference between this so-called 'winter-compensation' and the scheme described in the foregoing paragraphs is that the authority of T_3 must be increased.

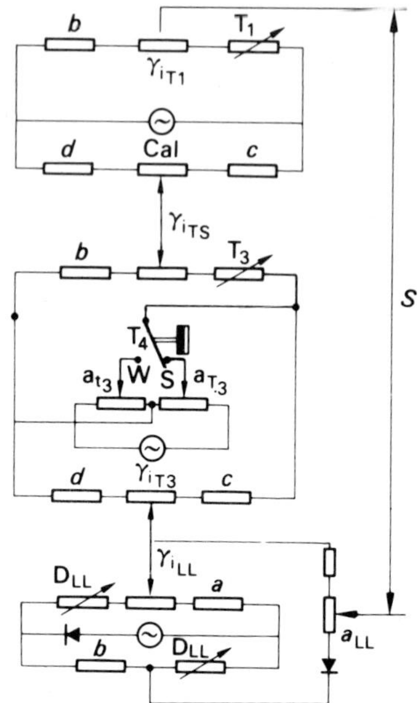

Fig. 5.17
Summer and winter compensation provided by a single outside sensing element T_3 by changing phase and magnitude of the auxiliary bridge voltage. A low-limit bridge is added for the sake of completeness. Symbols as in Fig. 5.15.

The opposite is done during summer for different reasons. The introduction of a dead zone between the heating and cooling stages allowed a limited rise in desired value as the weather gets warmer. The offset, during the cooling stage also causes the space temperature to rise above the desired value. Both these factors are uncontrolled and if increased to the extent required for reasons of comfort and economy alike, could not be relied upon to meet requirements under all circumstances.

It is therefore customary to reduce the dead zone to 1 – 2°C, to keep the proportional band during the chilling stage to a minimum consistent with stability, and to rely on a T_3 to raise the desired value as wished (Fig. 5.14 (*b*) and (*d*)).

5.6.3. Introduction of the T_3 signal

As is the case with 'electronic' limiting devices, the influence of T_3 does not of itself come to an end when its object is achieved. Thus, the sensing

(a)

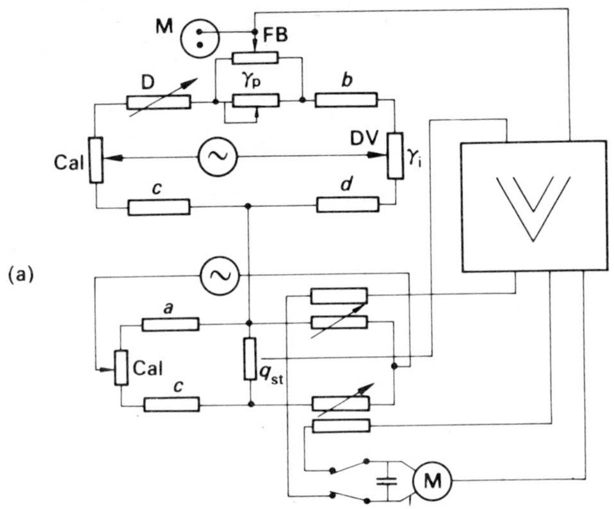

(b)

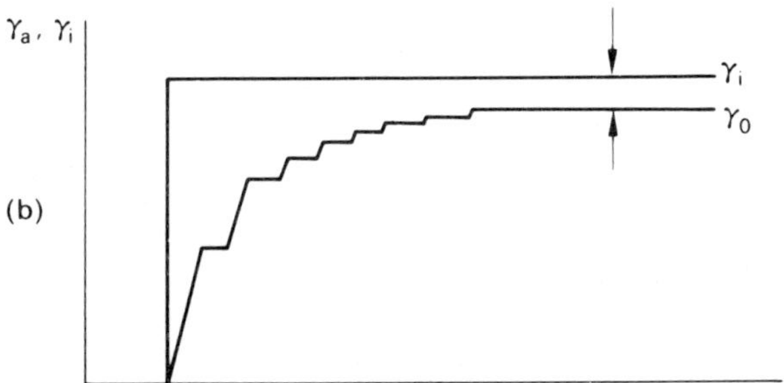

Fig. 5.18
P controllers with thermal feedback. (*a*) with separate bridges and adjustable stability bridge influence, a_{st}; (*b*) reaction of the controlled condition, γ_0, to a change in desired value, γ_i, with such controllers. Cal', balancing potentiometer of stability bridge.

element used to raise the desired value during the winter period would, unless disengaged depress γ_i during summer. In Fig. 5.15 a two-step thermostat T_4 switches over from the T_3 bridge for summer compensation to that incorporating a winter T_3. Owing to its larger control range the latter bridge also incorporates a snubber, T_2. Each sensing element has a separate adjustment for its desired value, γ_i and for its authority a. Owing to the differential introduced by the summer-winter thermostat T_4, the temperatures to be expected around the switch-over point vary slightly with rising or falling outside temperature. Figure 5.16 refers to a scheme without dead zone. In Fig. 5.17 feedback stabilising factors are omitted for the sake of simplicity. Interesting, however, is the way in which the same T_3 is made to provide summer or winter compensation as required by reversing the phase of the auxiliary bridge voltage; its authority during the heating and cooling phases being determined by the magnitude of that voltage in comparison with that of the main bridge. A low limit bridge is added just to show how further bridges can be tacked on as desired.

5.7. Stabilisation of P controllers by thermal feedback

The method as such has been discussed earlier in connection with floating controls (see Fig. 3.10). The stabilising effect is achieved by allowing a motor speed to increase proportionally to the magnitude of the deviation. Control and thermal feedback bridges may be combined, as shown earlier, or may be separate (Fig. 5.18(*a*)). The advantage of the latter arrangement is the adjustability of the stabilising signal to the control constants of the process. Also, small degrees of unbalance can readily be removed by means of potentiometer Cal' provided for the purpose.

Controllers of this type are sometimes offered as P + I controllers. As the bridge is rebalanced by a change in feedback potentiometer, on the occurrence of a deviation an offset will remain uncorrected. It is in this fact that the difference between the two types lies; no claims, other than that the stability of such P controllers is greater than that of the more common types, and that therefore the value of γ_P can be reduced, are justified in this respect.

6. Proportional controllers (Pneumatic versions)

6.1. General

The pros and cons of using compressed air to energise automatic control systems are detailed in Chapter 1. Though basically of the simple proportional mode (Fig. 6.1) pneumatic controls are equally available in I, P + I, P + D and P + I + D versions (derivative control is discussed in Chapter 8). By sufficiently reducing the proportional band and/or cutting out the feedback, controllers of the proportional mode respond as if their action were two-step. In addition, a large variety of amplifying, averaging, selecting, reversing and other relays have been developed. Therefore pneumatic control systems are as universally applicable as any of their counterparts. The fact also that they can be used safely in explosive atmospheres makes pneumatic control systems particularly attractive for use in many different industries. In large commercial air-conditioning plants, such as offices and other public or semi-public buildings, their popularity is based on price considerations.

The construction of pneumatic motors, whether piston or diaphram actuated (Figs. 6.1 and 4.2(*a*) respectively) is very simple, and their price is consequently so low in comparison to electric motors that, wherever valves and dampers are to be installed in appreciable numbers, pneumatic controls are the first choice, the savings on the actuators more than compensating the cost of the compressor and ancillary equipment.

If pneumatic controls are to be used to best advantage, a knowledge of the main features they have in common is essential – reason enough to devote a short chapter to the subject.

6.1.1. Types of pneumatic controllers

All pneumatic controllers convert a physical quantity, such as temperature, pressure, liquid level, humidity, position, or any other magnitude into a pneumatic signal, i.e. into an air pressure directly or inversely dependant upon the deviation. The difference between the

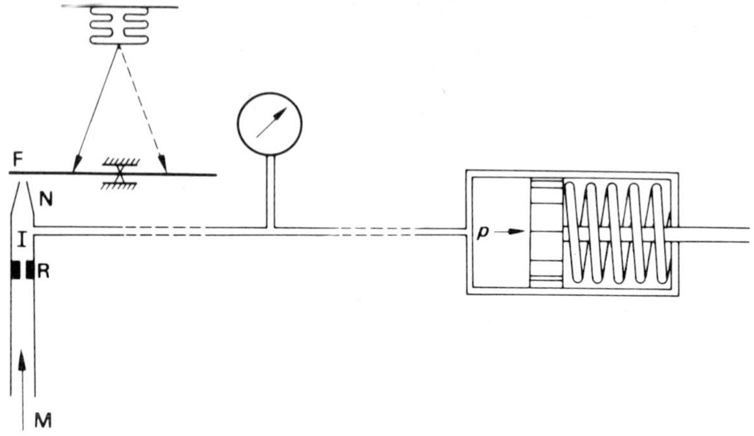

Fig. 6.1
Simple bleed-type temperature controller (DA solid line, RA dashed line). As flapper F closes nozzle N, the pressure in I rises and forces piston P down against spring tension.

simple bleed type (Fig. 6.1) and more sophisticated pneumatic controllers is that, in the former, the signal is applied to the actuator without further processing.

The bleed type has obvious shortcomings. All the air necessary to increase the pressure above the actuator diaphram and in the connecting pipeline has to pass through the restriction, R. When, subsequently, the pressure has to drop, the redundant air is exhausted through the nozzle N. The diameter of both R and N must be kept reasonably small in order to keep the air consumption down. Therefore, where large distances have to be bridged, the movement of the final control element may be too sluggish. A further important imperfection of this type is that, the application of negative feedback notwithstanding, the line showing the relation between controlled variable and air pressure, i.e. valve position, is curved (Fig. 6.2), and indicates a certain amount of hysteresis.

Both problems are tackled by the pilot-bleed system where, instead of using the whole of the standardised pressure range from 0 to 1·1 bar (0–15 p.s.i., both values relative to atmospheric pressure), the practically straight middle section of the characteristic of Fig. 6.2 is used. The pressure in chamber I monitors that in chamber II of a built-in amplifier

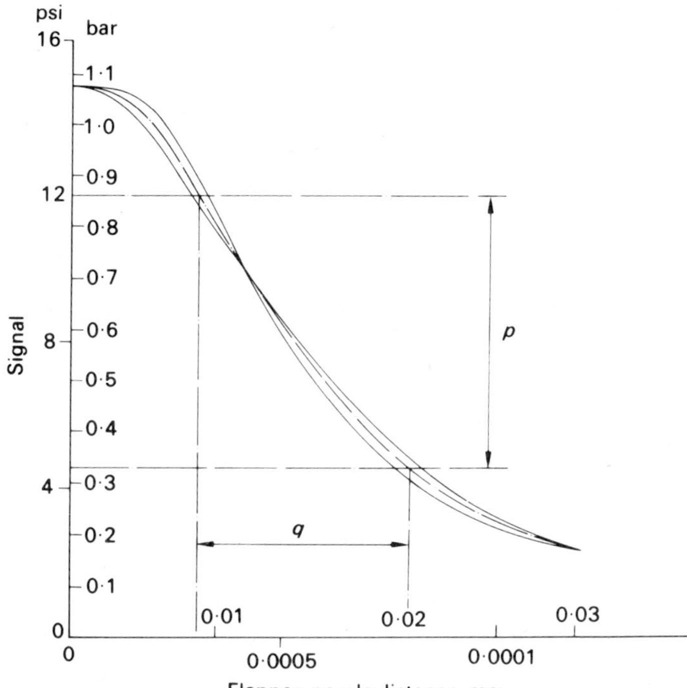

Fig. 6.2
Relation between input, i.e. flapper-nozzle distance, and output (air pressure relative to atmosphere) or bleed-type controllers.

which restores the original range of air pressures and increases the volume of signal air passing to the branch line B in as far as necessary (Fig. 6.3). Thus, though the restrictor and nozzle openings are considerably smaller, the valve reacts more quickly to a change in signal pressure and the linearity of the relationship between signal and γ_0 is improved. The hysteresis, though not wholly eliminated, is greatly reduced.

In this respect it is similar to the non-bleed system (Fig. 6.4) in which air consumption is theoretically restricted to the amount required to open or close the valve. In practice the ball valves in the controller are seldom wholly tight so that the air consumption is greater than the designation 'non'-bleed would suggest.

Transmitter or sensor schemes employ basically normal bleed type detectors, but only a tiny fraction of their (frequently enlarged) pressure range is used at any one time. The output signal of the detector

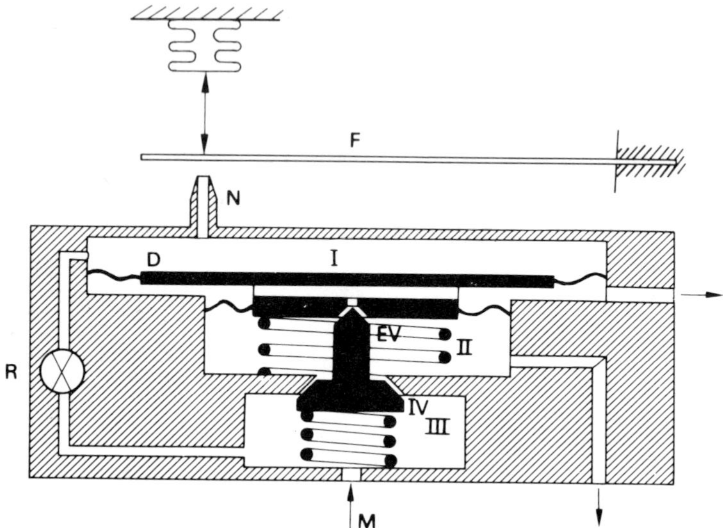

Fig. 6.3
Pilot-bleed pneumatic controller (DA). As N is closed by F the pressure in I rises. By pressing down disc D, it closes exhaust valve EV and opens inlet valve IV admitting main air from III to II as well as to branch line B until the force exerted on D restores the balance and closes both IV and EV.

is a measure of the magnitude of the controlled condition rather than of its deviation from the desired value, to which the signal is compared in a universal controller, i.e. one that can be used irrespective of the nature of the controlled condition. There, the deviation, if present, is established and amplified as required to re-position the final control element.

The main attraction of this system is that the signal can be used not only for purposes of control, but also to measure the magnitude of the controlled variable, the reading being indicated by means of simple suitably calibrated manometers. The same applies to the desired value, valve position or any other quantity that can conveniently be translated into an air pressure. Lack of linearity is tackled in a way similar to other types: by a judicious choice of the useful part of the characterictic an increase in mains pressure the application of negative feedback etc.

This classification of pneumatic controllers is by no means the only one in use, but the above characterisation of the various types will suffice for present purposes.

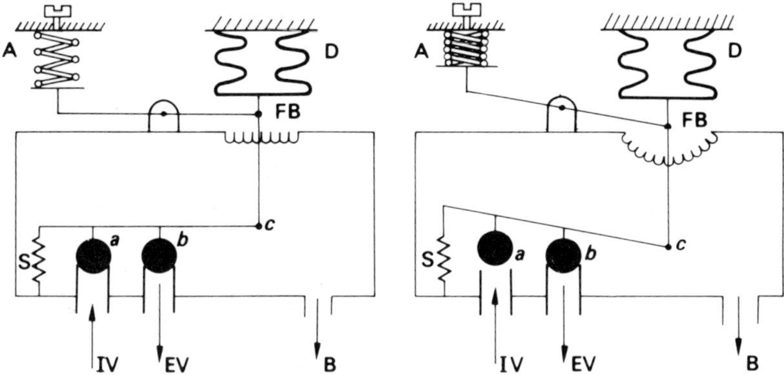

Fig. 6.4
Non-bleed controller (*a*) As bellows D expand *c* is depressed moving against the force of feedback diaphragm FB and of the two springs A and *s*, closing valve *b* and opening *a*. (*b*) Mains air increases pressure in branch line B, and forces FB back until balance is restored, and *a* closes once more.

6.2. Types of operation

Mechanically it is a simple matter to make the signal rise (direct action-DA) or fall (reverse action-RA) with an increase in controlled variable. The action of some detectors can be changed locally from direct to reverse action; with some other types the action is not reversible and has to be specified when ordering (Fig. 6.5).

A similar distinction can be applied to the actuator. Here, an increase in signal pressure must overcome the opposing force of a spring which, when the actuator is de-energised, causes the final control element to take up either of its two extreme positions (this spring-return feature inherent in the construction, by the way, is a further attractive feature of pneumatic controls especially useful when applied under conditions of potential danger). Whether in the absence of a signal, i.e. 'normally', the valve is fully open or wholly closed (whether a given actuator is n.o. or n.c. will depend on its type (Fig. 6.6), or, in the case of a damper, on the way the linkage is arranged. Some actuators can also be mounted upside down. Whereas normally an increase in signal pressure will cause the spindle to move further into the valve body, the reversed actuator has to withdraw the spindle against the force of the spring.

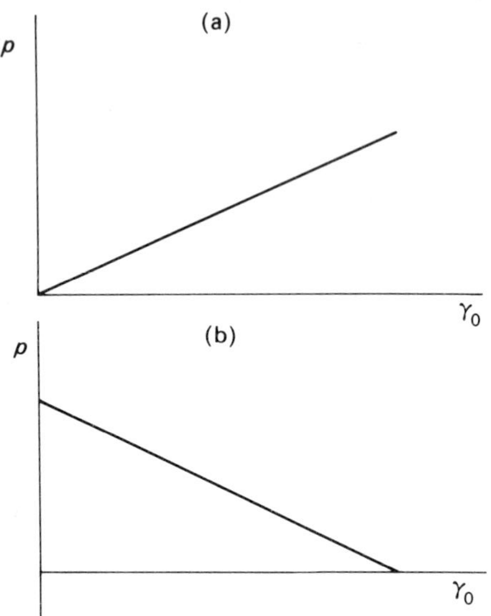

Fig. 6.5
When the signal pressure p increases with increasing variable γ_0 the action is called DA (a) when it falls it is called RA (b).

All this has an important bearing on the possibility of protecting the plant or process against certain hazards. Thus, if frost protection is of prime importance a heating valve will have to open and an outside air damper close should the air supply fail at a moment that danger is present. Therefore, the action of the respective actuators must be n.o. and n.c. and this, in turn, determines the action of the controllers; their signals will have to be DA and RA respectively if the valve is to close with rising temperature and the outside air damper to open up further. Should, on the other hand, a non-storage calorifier have to be protected against overheating, the valve would have to be n.c. and the controller action RA.

6.2.1. Further features

Let a simple control system, having come to rest, be subjected to, say, a step change in the desired value (Fig. 6.7(a)). The response of the actuator will initially be fast, as the difference in force exerted on the diaphragm by the signal pressure on the one hand will be large and

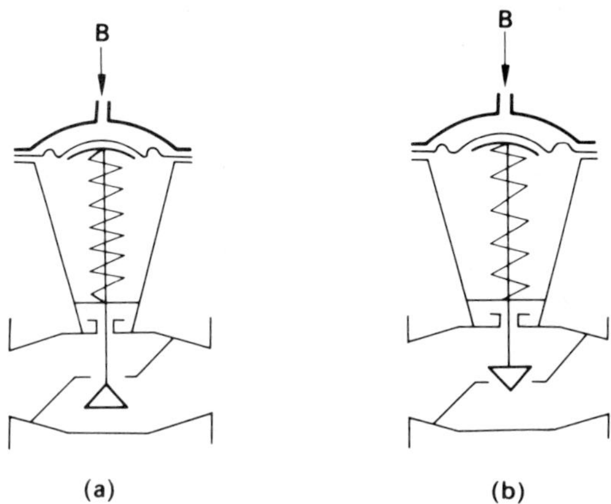

Fig. 6.6
Actuator-valve combinations that are (*a*) normally closed (n.c.) or (*b*) normally open (n.o.).

that by the spring, on the other, at a minimum value. As the valve approaches its new position, the spring tension increases and the valve movement slows down as the resultant force diminishes. The approach to the ultimate valve position is therefore slow, which, as has been mentioned in earlier chapters, reduces the chance of overshoot. This makes for greater stability than if the final control element were moved to its new position at constant speed, as is the case with the vast majority of electric motors. These are generally chosen to be of the squirrel-cage type on account of simplicity and cost and, so as to make up for the gradual approach of pneumatic actuators, are made to move at a far lower average speed, if one can speak of the 'average' speed of an actuator moving in accordance with an e-function.

A further interesting feature of pneumatic actuators is that by selecting a spring of suitable stiffness, the valve can be made to complete its stroke in response to a different range of signal pressures than the usual 0·2–1·1 bar (3–15 p.s.i.). Thus, if one valve is equipped with a spring that is fully compressed when the signal pressure has risen, say, from 2 to 8 p.s.i., and another that prevents movement below 9 p.s.i. and has completed its stroke at 15 p.s.i., the two valves can be made to move in sequence, in this case being separated by a dead zone equivalent to 1 p.s.i., expressed in terms of the controlled condition.

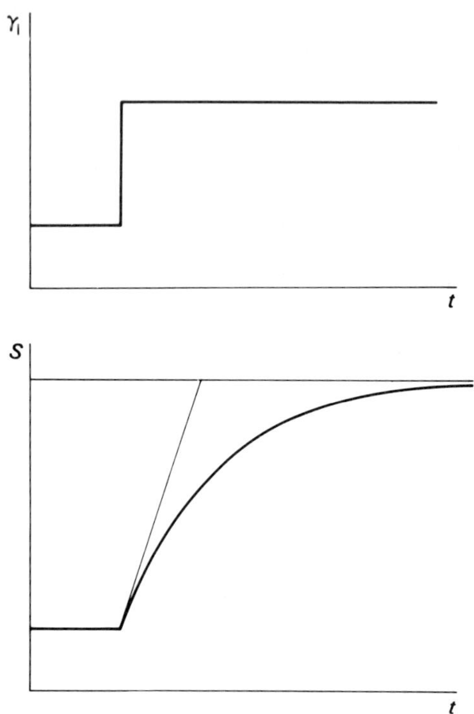

Fig. 6.7
The response of a pneumatic actuator to a step-change is initially fast, and slows down as the new position is approached, a valuable feature contributing to stability.

6.3. Valve positioners

The force necessary to move a valve against the differential pressure, increased under circumstances by the 'piston effect' (i.e. the effect of the static pressure forcing the spindle outwards), is, or can be, considerable. For a given signal, the force supplied by the actuator is directly proportional to the active surface area of the diaphragm or piston. It can be increased at reasonable cost and virtually at will by suitable choice of the actuator. Yet, usually, the actuator chosen will not be larger than is necessary in normal service.

In such cases, injudicious tightening of the gland may render movement jerky or even impossible, this leads to instability and introduces hysteresis which is equally undesirable. This has led to the development

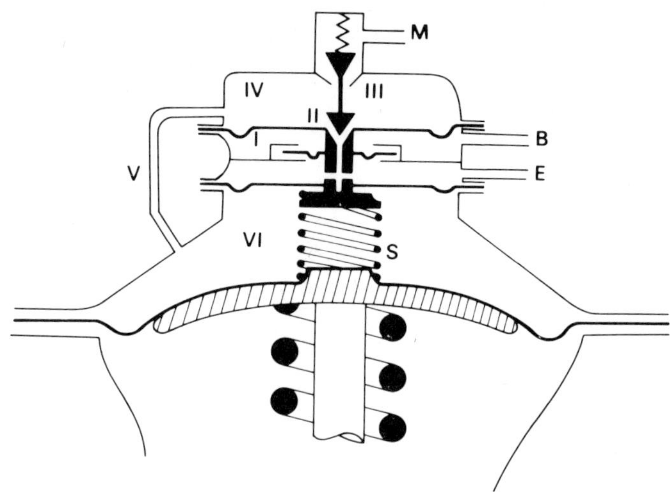

Fig. 6.8
A simple type of valve-positioning relay shown mounted on an actuator.

of the positive positioning relay, a device that has proved vastly more versatile than the above introduction would suggest. Whereas, without a positioner, when a signal smaller than normal does not produce the corresponding Δs because of friction, a positioner can draw upon the whole of the mains pressure to ensure that signal and valve movement are proportional at most times.

Figure 6.8 shows a simple type mounted directly on the actuator. The position of the control valve proper is conveyed to the central valve assembly II in the positioner by means of the force exerted by spring S. This force is assisted by that of the signal entering by branch B and which, when increasing, exerts a greater force upwards than downwards due to the difference in surface area of the upper and lower diaphragms enclosing the signal chamber, I. The valve seat, moving upwards, closes opening II, opens port III and admits air from the mains M at full pressure into chamber IV. Before the pressure in chamber IV can rise sufficiently to close valve III, that in chamber VI above the actuator diaphragm must assume the same value. As long as the main spindle is stationary the force exerted by the signal and the pressure of S, increased by the raised pressure in chamber VI, will keep valve II closed and valve III open. When at last the forces preventing movement of the spindle are overcome and it is forced downwards, the spring tension will drop, the valve mechanism will close port III and port II will remain closed: a new balance between signal and valve position is attained. It should be noted that the pressure in VI is primarily determined by the force necessary to overcome the friction increased by the differential pressure resisting movement of the valve spindle and disc attached to it. It therefore bears no firm relation to the magnitude of the signal. When the latter decreases, the central valve moves downwards, the pressure in VI is relieved via V, port II, and exhaust E, until the spindle moves upwards, increasing spring tension S and closing port II once more. Theoretically, the pressure in VI can be reduced to zero or increased to the full mains pressure should this be necessary to achieve the desired effect, irrespective of whatever value the signal may happen to assume at the time.

Positioners have far more applications than their name would imply:

(*a*) The volume of air required to raise the signal is reduced as only the pressure in the small chamber I needs to be changed. Moreover the air consumption of the positioner itself is restricted to the time during which the spindle has to move against the spring pressure.

(*b*) The speed with which the valve position is changed is increased owing to the greater range of pressures the positioner can draw upon to achieve its object, but the accuracy with which the final control element assumes its correct position is unimpaired.

(*c*) Positioners generally carry a special adjustment by means of which the magnitude of the signal necessary to move the valve over the whole of its stroke can be varied. Thus it provides a further possibility of adjusting the proportional band within wide limits.

(*d*) As a further consequence of this facility, the use of special actuator springs for the purpose of sequencing, as described in Section 6.2.1, can be avoided. This is a considerable advantage as their variety is necessarily restricted in number and when their marking has disappeared they are difficult to distinguish anyway. Theoretically the number of control valves operated in sequence by one single controller can be increased at will in this way. Also, their action can be made to overlap to any degree the process may require. Sequence control is frequently needed in air conditioning, where the heating valve must close before the outside air damper increases the amount of fresh air for cooling purposes. The dead zone that may have to be interposed between the ventilating and cooling stages can be effected by the same means.

6.4. Application of pneumatic P controllers

6.4.1. Cascade control

The simplicity with which any physical magnitude is converted into an air pressure has already been touched on. This fact facilitates the realisation by pneumatic controllers of a particularly useful scheme known as cascade, or master-submaster, control. Here the output of one controller, the master, is used to influence the set point of a second, the submaster. As this scheme is not restricted to pneumatics it will be dealt with in more detail in Chapter 10.

6.4.2. Remote control

So-called gradual switches are available with which a signal of any value up to full mains pressure can be initiated. These devices are used for remotely positioning valves and dampers and for resetting the control points of submaster sensing elements from a control panel or other central point.

A simultaneous change in a set value of a great number of sensing elements is frequently desired in large heating or air conditioning plants when, at the end of the working day, the space temperature is allowed to drop to a lower value for the night. Special pneumatic controllers have been developed for the purpose. By the simple means of changing the mains pressure to a value, significantly outside their normal range, say from 15 to 25 p.s.i. (1·1 to 1·75 bar), the set value is dropped by an adjustable amount. By the same means, other controllers can be made to change their action from DA to RA as, for example, may be desired in plants heated or cooled by varying the temperature of the flow to the same heat exchanger. This is the case in three-pipe induction units.

Enough has been said to show that pneumatic controllers have many interesting features in common and offer certain very real advantages. This should not be taken to mean that they are superior to other types in all ways. On the contrary, pneumatic detectors are frequently less sensitive and slower in response to a change in the controlled condition than electronic controllers; the necessity of providing clean and dry air for their actuation is a serious handicap for their use in smaller installations. The consequences of letting up on the quality of the air are so devastating that the process of drying, de-oiling and otherwise cleaning the air has to be carefully supervised. They are more sensitive to dust and other contamination than either electric or hydraulic controllers etc. Yet it is well worth considering in each case which type of controller offers the best results at the lowest price.

7. Proportional plus integral Control

7.1. General

P + I controllers are *two-term controllers*, combining the stability of the proportional and the accuracy of the integral mode. In this respect they can be considered a marked improvement when compared to the *single-term controllers* dealt with so far. The output signal of a P + I controller is given by the formula

$$V = -K_1\gamma - K_2 \int \gamma \mathrm{d}t$$

K_1 and K_2 are seldom chosen independently of one another; their ratio can be assigned the symbol

$$T = K_1/K_2$$

the significance of which will appear in due course.

7.1.1. The contradiction inherent in P + I action

It was shown in Chapter 4 how a purely proportional element was incapable of completely removing a deviation from the desired value: rather γ is cut down to a considerably smaller value, the offset, the existence of which was proven to be an essential pre-condition for proportional control action to take place. This means that a proportional output signal is inherently too small; if the accuracy of the integral action is to be achieved, a method must be found of increasing this signal without affecting the P control action. One way was described in Chapter 5: adding an open loop element, T_3, to the circuit can eliminate the offset to the extent that this is due to disturbances detectable by T_3. It has been shown, however, that open-loop elements, being unable to check the results of their own action, must necessarily fall short of the ideal, especially as they are not subjected to all disturbances experienced by the controlled variable itself. However, the

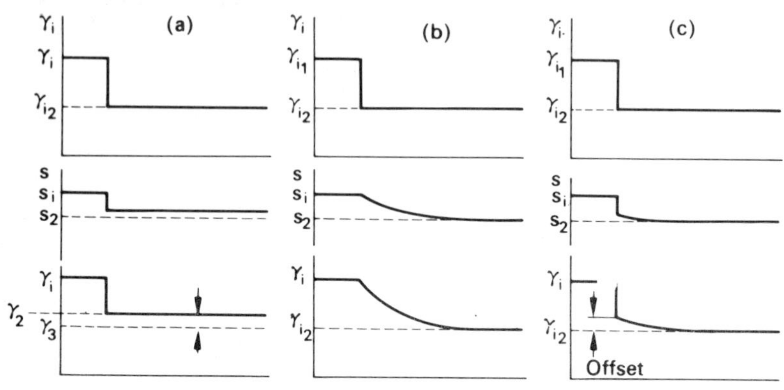

Fig. 7.1
The response to a step-change of (*a*) P controller; (*b*) I controller; (*c*) P + I controller $T_m = \tau = T_{de} = 0$

alternative, i.e. P + I control action achieved by two closed-loop elements working independently of one another, is no less problematical.

Assume, for instance that the P element, attacking a deviation first, has reduced it to the offset. The I element, recognising the offset as a departure from the desired value, will initiate further action to eliminate it. In moving the final control element further in the same direction, the balance between the bridge output and feedback signals is disturbed: the P element will strive to being about the condition existing before the I element interfered. This inherent contradiction must be resolved if the two elements are to be prevented from warring continuously.

7.1.2. The contradiction resolved

Leaving this problem for the moment Fig. 7.1, (*a*), (*b*) and (*c*) show schematically the reaction of ideal P, I, and P + I controllers to a step change in circuits with neither exponential nor any other type of lag.

On a change in desired value a P controller will remove the deviation short of the remaining offset with a speed limited only by the response of its detector and that of the actuator. An I controller will cause the new value to be approached in accordance with an e-curve and at an initial speed dependent upon its adjustment. A P + I controller, finally, will combine both actions; in this simplified representation, the action of its I element is limited to the removal of the offset.

Figure 7.2 recognises the difficulty described in the previous section and shows one way of solving the problem. Let the control circuit have

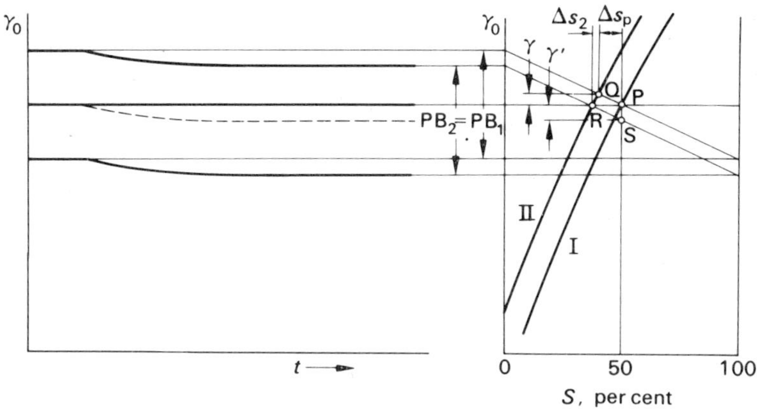

Fig. 7.2
The action of the I element of a P + I controller explained as an automatic shift in set value sufficient to keep the controlled variable constant at the desired value.

come to rest at point P where $\gamma_o = \gamma_i$ and γ is therefore equal to zero. If the load is then reduced, the process control characteristic will move upwards. The point at which equilibrium exists between γ and s shifts from P to Q, leaving the offset in the first instance uncorrected. If it is now supposed that it were not the function of the I element to take any direct control action but to work indirectly by shifting the desired value slowly downwards until the control point, sliding down along the p.c.c. reaches the level of point P, the slight change in valve position, Δs_I, necessary to remove the offset is effected. No difficulties arise from the P element since, far from the offset having been removed, the difference in the value of the controlled variable at points R and S has, due to the curvature in p.c.c., even increased when compared with that between points P and Q.

In this representation of P + I action, the proportional band is supposed to move upwards and downwards with, and symmetrical to γ; as far as is necessary to keep the control point at the same value. As, once more, the supposed shift in γ_i takes place exponentially the complete removal of γ will, in theory, take infinitely long, independent of the way in which the controller is adjusted.

7.1.3. Integral action time

This purely theoretical point of view apart, wide variations in the speed with which the I component makes itself felt are encountered in practice. Therefore, it is obviously useful to define an integral action time as a measure of the vigour of the integral control element. Figure 7.3 refers to a broken-loop system with no lags whatsoever ($\tau = T_{de} = 0$). Now the integral action time, T_I, will have elapsed when the individual responses to a step change of the P and I elements are equal.

This definition may seem rather contrived and prompted by theoretical considerations only. Yet, however little relevance to control practice it may have, if only on account of the supposed absence of lags, the continuously changing conditions in a closed-loop circuit as well as the interaction of the many factors involved make if difficult to devise a better one. In actual practice the response of both elements will reduce γ and thereby their own output signals: the straight lines of Fig. 7.3 become curves. It might be more meaningful to say that the integral action time is the time the I element would need to move the valve over the same distance as the P element, assuming both initial responses to remain unchanged. This definition is reminiscent of that of the time constant given in Section 1.3.2. The inference that the integral action time corresponds to the time constant of the I response curve can be confirmed theoretically. Furthermore, it follows from the standard definition that if

$$-K_1\gamma = -K_2 \int_0^T \gamma \mathrm{d}t = -K_2\gamma T$$

the integral action time equals

$$T = \frac{K_1}{K_2}$$

Figure 7.3 clearly illustrates that the action of a P + I controller should not be looked upon as the mathematical sum of that of a P element and of an independent I element. At (*c*) the initial proportional band is halved as a result of which the rate of change of the I response is doubled. When, as at (*d*), γ_P is doubled, the I output signal changes to half the original speed. Yet in both cases the integral action time has remained the same. Clearly, the response of the I is greatly influenced by the proportional band to which the controller is adjusted. The controller feedback, characteristic of a P controller, is both the latter's strong and its weak point. On the one hand it limits the valve move-

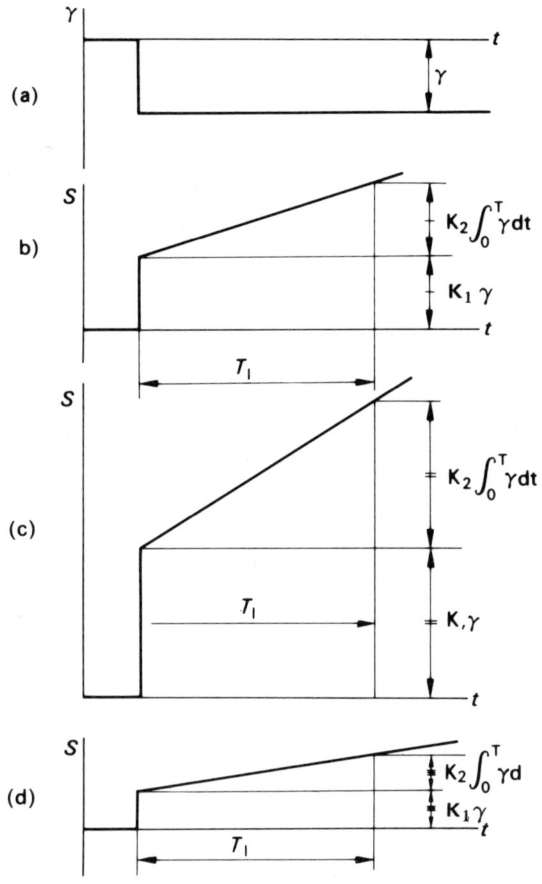

Fig. 7.3
The concept of integral action time explained by the reaction to a step change of the P and I elements individually in a broken-loop system with no lags whatever. (*a*) a deviation appears; (*b*) response of the P element, $K_1 \gamma$ and of the I element, $K_2 \int \gamma dt$. When the two are equal the integral action time, T_I, has elapsed; (*c*) response of both elements when γ_P is halved; (*d*) response of both elements when γ_P is doubled.

ment for a given deviation which, with I controllers, is subject only to the physical limitation of the valve stroke. Thus, the inherent tendency towards instability, so characteristic of integral action, is avoided by P controllers. On the other hand, this very feature is responsible for the offset. In a P + I controller, therefore, an arrangement had been devised on which the feed back signal appears in full strength every time the valve position is changed and subsequently seems to die away. Thus the offset is eliminated without causing instability and the results are considerably better than those of either an I or a P controller. This accurately describes the P + I control action, and is the reason why, P and P + I controllers are respectively called controllers with rigid and with yielding feedback in German nomenclature.

7.2. The influence of the control constants

Bearing in mind that the integral action time T_I, in itself is no direct measure of the vigour with which the I element reacts to a deviation, it is permissible, when surveying the response of a P + I controller in general terms, to consider the response of the two elements to the process characteristics separately. The tolerance of a P controller of a fairly long T_{de}, provided τ is correspondingly large, would allow γ_P to be fairly narrow in P + I control. However, the effect of this on the I action must still be checked. If under these conditions a tendency to oscillate is noticed at the largest value of T_I the I element will necessitate an increase of γ_P. When, on the other hand both T_{de} and τ are short but T_{de}/τ is small, the P element may require a wide γ_P, whereas the integral action time can be shortened.

Needless to say, a large control range is unfavourable to both elements. As τ decreases, so that the process responds rapidly to both disturbance and the corrective action it invites, the integral action time must be correspondingly short if a P + I controller is to be of any real use. Thus, when milk is being pasteurised by passing it over the surface of a non-storage calorifier type of heat exchanger, T_I may have to be of the order of seconds: when keeping the frequency of a generator output constant, it will be of the order of milliseconds and less.

In many heating and air-conditioning processes, on the other hand, dead time and similar lags may be fairly long, and T_I may well have to be increased beyond what many P + I controllers can offer. In general, however, the adjustment is in no way critical. Fear of a large offset as a result of increasing γ_P for the sake of stability loses much of its point in the knowledge that the I element will eventually eliminate it, especially as T_I can be reduced under these circumstances.

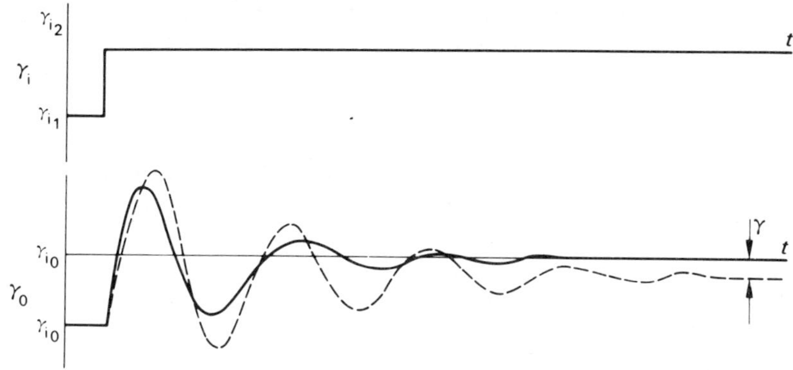

Fig. 7.4
P + I response to a change in set value γ_i (full line) compared to the (broken) response of a purely proportional element which leaves the offset γ uncorrected.

7.3. The real nature of P + I response

In industrial processes, I times varying from a fraction of a second to the larger part of an hour may be required; in the heating and air-conditioning field I times less than, say, 5 min are rare. So as to remove any lingering suspicion that, in order to remove the offset, it will take the full five, fifteen, thirty or to whatever number of minutes, T_I has been adjusted, the subject, already touched upon several times, must be raised once more, at the risk of being thought repetitious. This question is as far removed from reality as the official definition of T_I that prompted it. Assuming for the moment that I action is withheld until the P element has reduced the potential deviation to the offset subsequently to be removed by the I element, the time the latter will need will be roughly in proportion to the magnitude of its control action. A circuit in which a P controller cannot reduce a potential deviation by more than factor 4 (i.e. $G = 3$, see Chapter 4) is a reasonably difficult case. This means that, assuming the p.c.c. to be a straight line, the valve movement will be in proportion to the action of the two elements, so that the time required to remove the offset is likely to be at most a quarter of the time a purely I controller would need to remove the whole of the deviation. Since the speed with which the I element moves the valve is proportional to the deviation, the first three quarters or even seven eighths of the offset will be removed more quickly than the rest, which is therefore rapidly reduced to a value of little importance. Actually, neither point, however obvious it may seem, is relevant. The two elements respond simultaneously instead of in sequence. The representation of Fig. 7.4. shows as a broken line the response of the P element alone to a change in set value; the full line shows the P + I response, the difference the two being the influence of the I element.

The moment γ_i is changed, both elements come into action, the I output signal reinforcing that of the P. Since the P + I signal is greater than either of its components, the new desired value, γ_{i_2} will be reached earlier than if the P element alone were active. At that point the I signal will have disappeared. Usually the valve will move further than necessary: as γ_o overshoots the mark the I signal will tend to reduce this excessive valve movement. Both elements then join forces so as to keep the growing deviation within limits; as a result, the overshoot is less and the valve is returned more quickly to its correct position than if only the P element were active, as shown by the broken line, and so on. Based on this figure, the view seems justified that the I time is largely irrelevant. One might say that, in bringing γ_o back to γ_i, the P element has done all the work, the function of the I element being reduced to preventing γ_o from departing from the desired value it has already reached.

Figure 7.2, on the other hand, suggested that the I component was mainly responsible for the control action: in simulating a shift of γ_i from point *P* to *S*, it forced the P element to move the valve as the circumstances dictated.

Each view represents a very real aspect of P + I response, and whichever one is preferred, they both demonstrate how confusing the concept of the integral action time must be.

7.4. Some P + I controllers

The hydraulic P + I controller of Fig. 7.5 differs from the P version of Fig. 4.8(*b*) inasmuch as a flexible element 11, 12 and 13 has been placed in feedback rod 8. Needle valve 13 provides a bypass over piston 12 by means of which the flow from one side to the other can be accurately adjusted. When, at *C*, a change in γ_o is detected, point 2 will act as a fixed point around which *A* and *B* revolve. The valve mechanism 5, 6, 7 admits pressure to one or other side of piston 9, and in moving it changes the position of final control element 10. The aperture 13 being small, the flexible link acts as if it were rigid; point 2 moves in

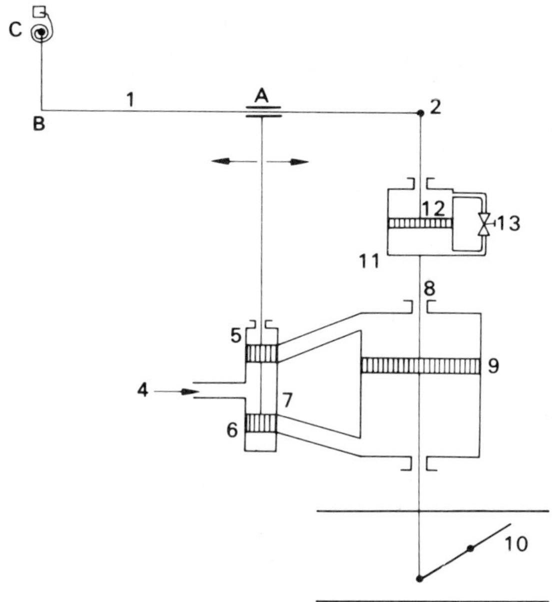

Fig. 7.5
Hydraulic P + I controller.

sympathy with 9 and 10 and readjusts valve 5, 6, 7 to its mid-position. However, as the fluid in cylinder 11 is under greater pressure on one side of piston 12 than on the other, it starts to flow through 13, thereby modifying the effective length of rod 8, and causing points 2 and *A* to revolve around the fixed point *B*. This results in a further readjustment of the final control element in the same direction as before.

The basic principle of the pneumatic P + I controller of Fig. 7.6 is very similar. The bellows placed in the feedback connection is opposed by the top bellows which acts as if it were rigid to a sudden impulse from the sensing element but as the air seeps through the variable restriction R, by which the integral time is adjusted, the feedback signal dies away.

In electronic P + I controllers, the essential part is an *RC* unit (Fig. 7.7). The (d.c.) feedback signal is applied to a capacitor which, in charging up, transmits it practically unchanged to the next stage where it limits the action of the signal in the ordinary way. The capacitor is bridged by a resistance through which the charge gradually leaks away, thus once more causing the feedback signal to die away in time. The integral action time attainable is limited by the capacitance of the

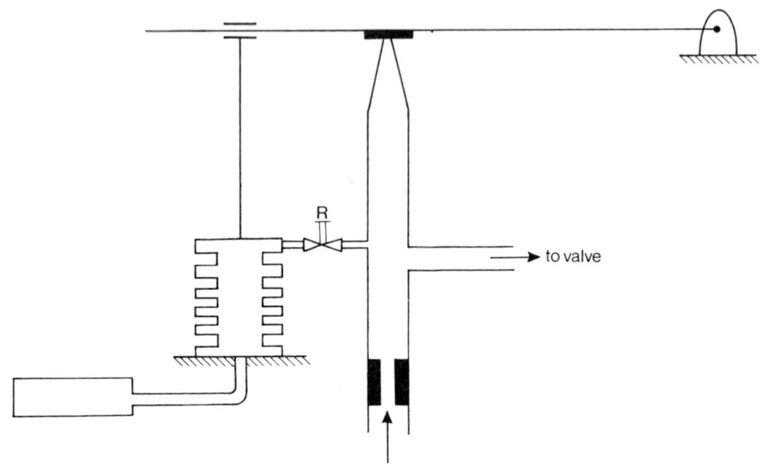

Fig. 7.6
Pneumatic P + I controller.

capacitor and its inherent leakage. 250 μF is a very high value for a capacitor of reasonable size as is 6 megohms for a resistance: The time constant of such an arrangement, assuming the leakage current in the condenser to be negligible, equals

$$T_I = \tau_I = RC = \frac{250 \times 10^{-6} \times 6 \times 10^6}{60} = 25 \text{ min}$$

Normally, however, the leakage is by no means negligible: the internal resistance R' provides a path parallel to R thus reducing τ_I in the ratio

$$R: \frac{RR'}{R+R'} \quad \text{or of } 1 : \frac{R'}{R+R'}$$

7.5. The block diagram

Figure 7.8 shows the action of the elements that, together, form the control loop. As usual, the process is characterised by its τ, T_{de} and Sp; the bridge, of which the sensing element is a part, initiates a signal θ

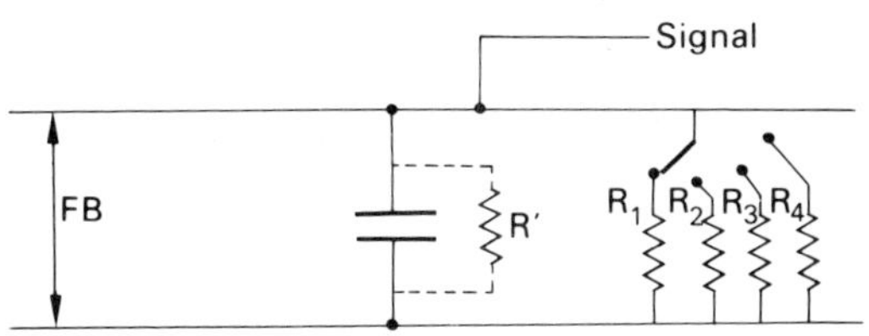

Fig. 7.7
The heart of an electronic P + I controller: an *RC* unit, FB, a feedback signal.

proportional to γ. This signal in its turn applied to a P + I controller generates a signal characterised by T_I and K_1, which causes the (electric) motor to travel over a corresponsing part of its stroke s. The resulting action, A of the controlled device is fed back into the process.

7.6. Practical application of P + I controls

7.6.1. Adjustment

Instructions for the optimum values of γ_P and T_I have been published by several workers in the field.[4,12,14] They are generally derived from experiments and, as the conditions upon which these were based were not necessarily exactly alike, may lead to slightly different results.

Far from casting doubt upon the accuracy of such instructions, this confirms that the adjustment of a P + I controller is not usually critical. Frequently the adjustment must be carried out even though the characteristics of the process are both unknown and unobtainable. In this case T_I is adjusted to its maximum value or, if possible, cut out altogether so that the P + I controller reacts as if it were a single-term P controller. A deviation is introduced, e.g. by changing γ_i a little, and γ_P is reduced until the loop starts to oscillate. The critical value of the proportional band (PB) is noted and the period of the oscillation T_{cr} timed (Fig. 7.9). The PB is then doubled and T_I adjusted to approximately 0·85 times T_{cr} (method 1, Ziegler and Nichols[14]). When the response to a step change can be taken, the control loop being broken as described earlier, the second method developed by Ziegler and Nichols applies (Fig. 7.10):

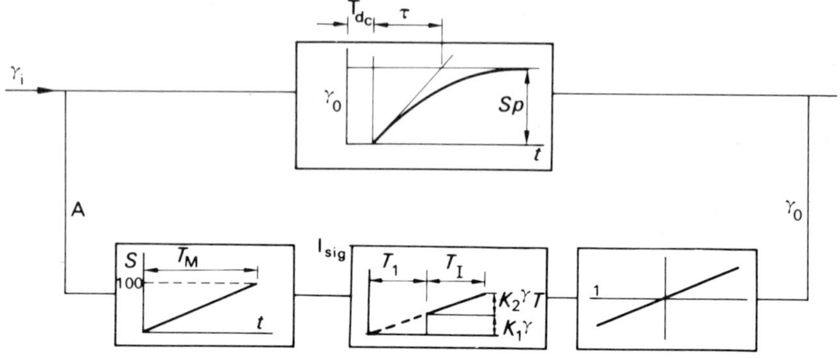

Fig. 7.8
Block diagram of a control loop containing a P + I controller.

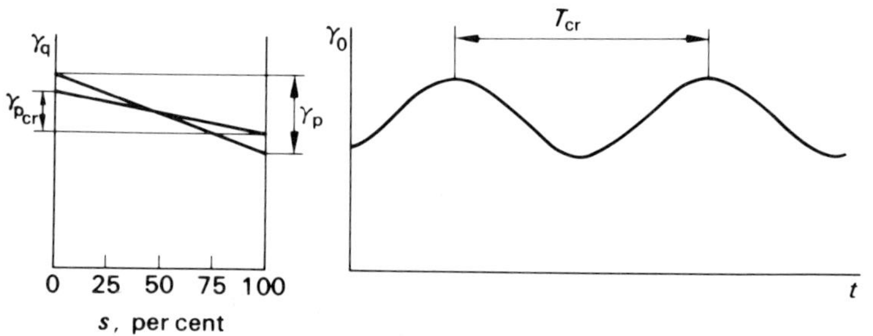

Fig. 7.9
Adjustment of a P + I controller by experimental means according to Ziegler and Nichols[14] method 1.

$$T_I = 3{\cdot}3L \text{ in minutes}$$

$$\gamma_P = 110\frac{K}{M}$$

in which M is expressed as a percentage of the span, and K in the units of the controlled variable. When the curve is taken for the case that the valve is made to run through the whole of its stroke, the process amplification equals

$$a_p = \frac{100N}{M}$$

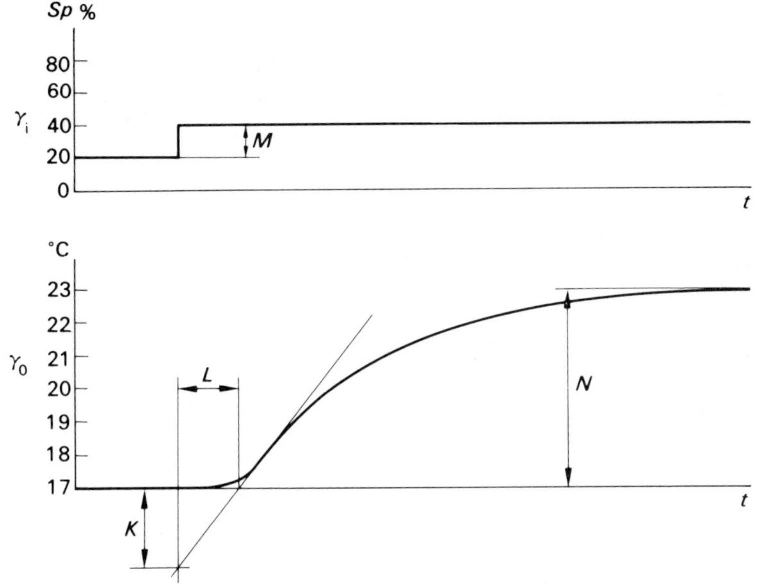

Fig. 7.10
Determination of the optimum adjustment of a P + I controller with the help of the response to a step change according to Ziegler and Nichols[14] method 2.

The controlled amplification equals

$$K_{c_1} = \frac{M}{1{\cdot}1\,K}$$

Chien, Hrones and Reswick[4] as quoted by Oppelt published the conditions for the most rapid aperiodic response to a step change as

$$\gamma_P = 2{\cdot}9\,\frac{T_{de}}{\tau}$$

and

$$T_I = 1{\cdot}2\,T_{de}$$

A P + I controller can oscillate as well as any other. The oscillating frequency will depend both on the process constants and on the element

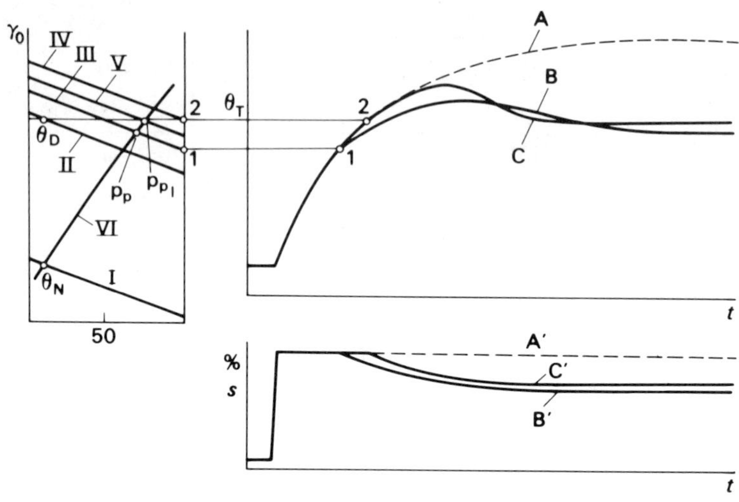

Fig. 7.11
The tendency to overshoot of P and P + I controllers compared. A, A′ without, B, B′ with P control, C, C′ with P + I control.

causing the trouble. If the oscillation period is relatively long, γ_P should be widened., after which the I time can possibly be shortened. A more rapid oscillation generally indicates that the I time should be increased.

7.6.2. The P + I response to a large step change

A controller with proportional plus integral action is the most sophisticated type of controller dealt with so far. However, it has one shortcoming not immediately apparent, namely its tendency to over- or undershoot, when, controlling a process with considerable lags, a disturbance or change of setpoint necessitates an appreciable displacement of the final control element. It can also occur when starting up, and this is the reason why, in industrial processes, a P + I controller is generally not switched on until manual control has brought the controlled condition to a value close to γ_i. The reason can conveniently be explained with the help of a further example from the heating field, the night set-back, or rather the restoration of the day control point early in the morning (Fig. 7.11) the undershoot during the night which is equally likely to occur, being of minor interest.

After having settled down at the night temperature θ_N on line I, under load conditions represented by p.c.c. VI, the desired value is suddenly raised to θ_D, the desired value during day-time. The valve will immediately open up completely as γ_o is well below the lowest point of the new controller characteristic II. As the process is supposed to be characterised by considerable lags of various kinds, the temperature rises slowly. Under these conditions a single term P controller would be characterised by line III and would start to close the valve as soon as γ had reached point 1. This would reduce the rate of rise, and although in any other than aperiodic response, some overshoot is inevitable, it is limited by the closure of the valve as the new control point P_p is approached.

The I element of a P + I controller, on the other hand, in trying unsuccessfully to reduce the suddenly enlarged deviation by further opening the already wholly opened valve, will slowly shift the control line upwards. By the time point 1 is reached, line II has fetched up at position IV. Action is delayed until the new desired value is reached at point 2; the overshoot is correspondingly larger than in the case of a simple P controller.

From what has been said, the methods by which the effect of this characteristic can be reduced will be obvious. They consist of avoiding situations in which large valve movements are necessary and, if this is not possible, increasing the integral action time as much as conditions will allow.

8. Derivative control

8.1. General

A survey of the primary modes of control would not be complete without reference to derivative control even though it is invariably applied in combination with one or both of the modes dealt with in preceding chapters. B.S. 1523 describes derivative action as that of a control element whose output signal is proportional to the rate at which its input signal is changing. No reference is made to the desired value; the signal of a derivative element will be the same whether the deviation is large or small. As, therefore, a purely derivative controller could not be relied upon to return the controlled variable to the desired value, this essential function must be carried out by other means. Combining a D element with a P element alone, or together with an I element, one obtains a P + D or a P + I + D controller, two- and three-term controllers respectively.

Returning to the D element itself, its action can be expressed by

$$V_D = s_2 - s_1 = -K_3 \frac{d\gamma}{dt}$$

in which K_3 is the derivative action factor usually expressed in terms of the action factor of the element(s) with which it is combined. Thus, the derivative action time of a P + D controller reacting to a variable of which the deviation changes at a constant rate, equals the time interval in which the part of the signal due to the P action increases by an amount equal to the part due to derivative action.

8.2. The action of a D element

Figure 8.1 illustrates what is meant by this rather involved definition. Starting from a condition in which the deviation remains constant,

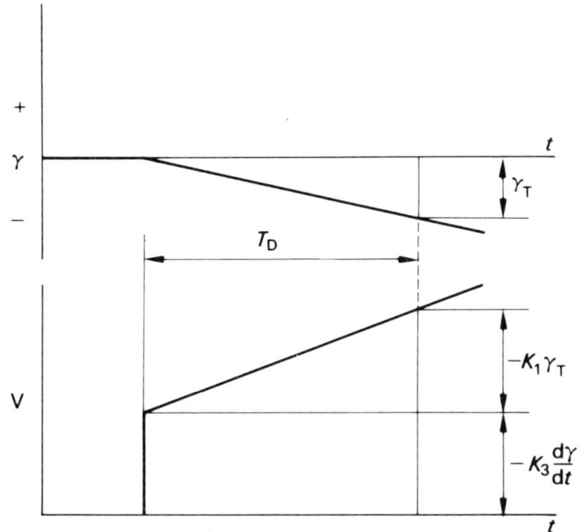

Fig. 8.1
The derivative action of a P + D controller will have elapsed when in a controlled condition changing continuously at a steady rate the action of the proportional element equals that of the derivative element.

neither the P nor the D element changes its signal. Assume, however, that at a given moment γ starts to increase at a constant rate $d\gamma/dt$. This causes the D element to emit a signal proportional to that rate of change:

$$V_D = K_3 \frac{d\gamma}{dt}$$

The output signal of the P element, starting from zero, increases in proportion to γ:

$$V_P = -K_1 \gamma$$

When the two output signals are equal, the derivative action time T_D has elapsed. The fact, that γ continues to increase notwithstanding the fact that the signal of both elements should normally have effected the slope shows that the closed loop has been cut at some point, e.g. by disconnecting the motor, or by closing a cock in series with the control valve. The definition as such, therefore, has as little relevance to normal practice as that of the integral time had.

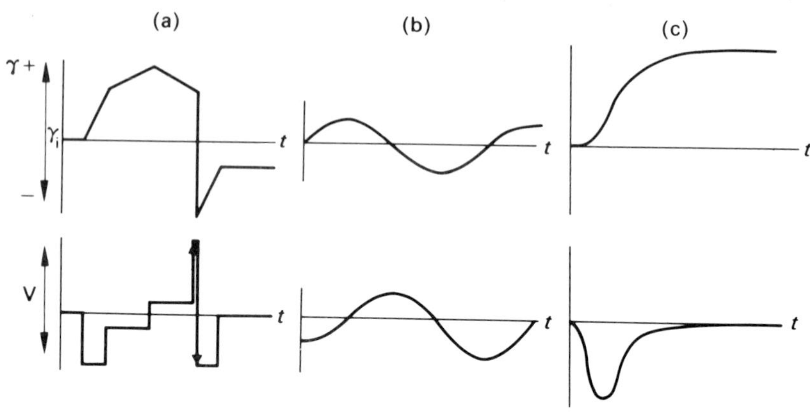

Fig. 8.2
Derivative action demonstrated as a reaction to arbitrary changes in the controlled condition (*a*) Reaction to γ_o changing continuously at several different rates of change and in different directions; (*b*) When the input signal is a sine wave, the output of a D element will vary a cosine wave; (*c*) D reaction to a typical higher order response curve;

As was explained in the chapter on the P + I controller, the effect of the D element is not an independent entity, but depends on the gain, i.e. on the proportional band, to which the P-element is adjusted. Reducing γ_p will increase $K_1\gamma$, so that even though both the derivative action time and the rate of change of the controlled remain the same, K_3 will also increase; the derivative action becomes more vigorous. If it is suspected that a tendency to become unstable is due to the D element, stability may be restored by increasing γ_p as well as by increasing T_D. Fig. 8.2 in which the deviation is changed arbitrarily in order further to clarify the D action. As long as γ does not change, V remains constant. As soon, however, as γ starts to move at a certain speed, an output signal appears that will cause the valve to move over a distance wholly unaffected by the present or future magnitude of γ. Only when the speed is reduced to one third, the valve displacement is reduced in proportion even though γ continues to grow. The rate of change then changes its sign, as does V. A sudden drop in γ causes the valve to open completely only to change sign when $d\gamma/dt$ reverts to its original value, even though γ is now being reduced the valve returns to the position it took up after the first change. Finally, when γ once more becomes constant, a D element on its own will return the valve to its mid position, however large the remaining deviation may be.

At (*b*) it is shown how a sine-shaped input signal is transformed into a cosine output by the D element. The reason will be obvious to the reader who has followed the explanation to Fig. 8.2(*a*) The two curves are identical but for the phase difference of $\pi/2$ by which V_o leads with respect to $d\gamma/dt$. This has earned D action the reputation of being able to anticipate the final value of a change provided, of course, that the change has started to manifest itself.

In processes with a pronounced transfer lag, such as that shown at (*c*), the anticipating nature of a D element is shown in a different way. One gets the impression that a D element deduces from the rapid rate of change at the osculation point that a large deviation is about to develop. It therefore initiates an appreciable change in valve position at a moment when the output signals of both an I and a P element would still be small. Should γ continue to grow, albeit at a reduced rate, the D element will gradually disappear, leaving the P or I elements to carry on the good work.

These characteristics are of great value where, in a process of higher order, it is essential that the variable should be maintained at the desired value with a high degree of accuracy.

8.3. A simple D element for temperature control

Though there are plenty of processes with pronounced transfer lag in the heating field, e.g. floor heating and the simplest of all control methods, a room-stat controlling an oil-burner, these are not applications where a high control quality is needed; the controller must be as cheap as possible. In air-conditioning processes, transfer lag is seldom so pronounced that special measures are required to counteract it. Yet some control firms can provide a reasonable approximation to this facility for electronic temperature controllers at a relatively low price. Of two identical wire-wound sensing elements, one is covered with a layer of insulating material, and the other is directly exposed to changes in ambient temperature. The two sensing elements are wired up in an auxiliary bridge circuit as shown in Fig. 8.3(*a*). A step change in temperature only calls forth a reaction from the bare winding and the outputs signal will be in accordance. Gradually, however, the change will penetrate the layer of insulating material and the temperature of the second sensing element winding will also start to change. As it approaches the temperature of the first element the bridge output will

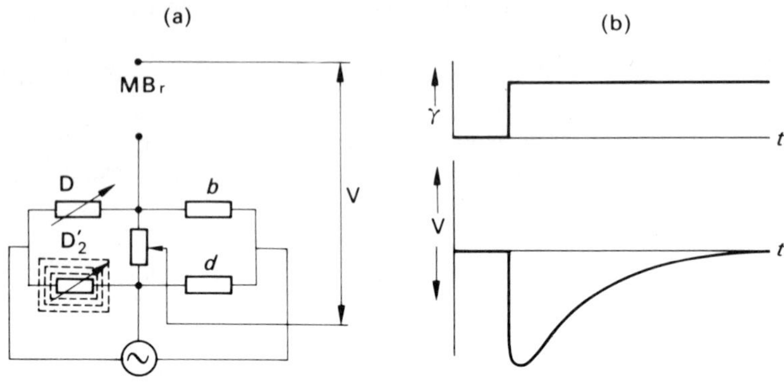

Fig. 8.3
(*a*) A simple D element D_2, D'_2, linked up in a bridge circuit and providing a D input signal for an electronic controller; (*b*) Change in valve position when controlled solely by a D-element as at (*a*) following step change in desired value γ_i.

diminish progressively. The two detecting elements being identical, the two opposing signals will eventually cause V to disappear altogether.

The greater the rate of change of γ, the greater the lag with which D'_2 follows D_2, and the greater the output signal. Inasmuch as V dies away gradually instead of dropping back to zero as soon as γ has become constant, the signal provided by this simple device is but an approximation of a true D element.

The same principle is used to improve the action of the dual and triple thermostats described in Chapter 5. The snubber thermostat T_2 is covered by a layer of insulation over which a second winding, identical to the first, is wound. Its output signal when used as a snubber is in accordance with Fig. 8.3(*b*); as the signal dies away in time one can afford to increase its authority with respect to the signal from the controlling thermostat without the offset being materially affected by its proportional band. The γ_P of T_1 can be reduced even more than in a normal two- or three-thermostat scheme.

8.4. Combining a derivative element with other elements

It has already been made clear that a D element can be very useful in taking action of the right kind at a time that no other control element would do very much. Thus it prevents the deviation from developing to the extent it would in its absence.

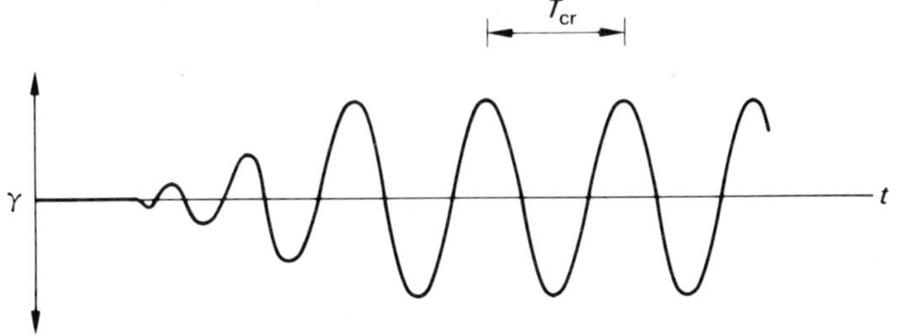

Fig. 8.4
The proportional band is reduced until the system starts to oscillate with constant amplitude. The oscillation time T_{cr} is then noted together with γ_P.

Yet, it will also be clear that the influence of a D element must be severely limited; if not, it may easily move the final control element to such an extent, that a deviation of opposite sign is developed equal to or larger than the one that gave rise to the original signal; there is considerable danger of instability. Furthermore, the inherent characteristics of the element(s), the action of which the D element is to supplement, are unchanged: P elements retain their offset and I elements continue to tend to instability. The latter shortcoming is even enhanced by the addition of the D element. This is the reason why P + D controllers have their use in industry, whereas the combination I + D is rarely if ever encountered.

P + I + D controllers offer the most comprehensive method of control to be dealt with in this book: the P element offers the stability, the D element restricts the deviation and the I element removes the offset inherent in the proportional system.

8.5. Adjustment

The two American control engineers Ziegler and Nichols, already referred to in the previous chapter, have published a method of adjusting a P + I + D controller to best advantage bearing in mind the characteristics of the process it is to control. Both the I and D action knobs are turned to their maximum value and the proportional band is increased to ensure stable operation. Then the proportional band is reduced with small steps. It will take longer and longer before the oscillations die down after a disturbance, e.g. a small change in desired value, has been introduced. At a certain value of proportional band γ_{Pcr} the system will no longer come to rest, but will continue to oscillate at constant amplitude and frequency $1/T_{cr}$ (Fig. 8.4). The control is then adjusted to the following approximate values. The proportional band to $1{\cdot}7\ \gamma_{Pcr}$, the integral action time to $T_{cr}/2$; and the derivative action time to $0{\cdot}12\ T_{cr}$.

Chien, Chrones and Reswick derive details of the optimum adjustment from the response to a step change:

$$\gamma_P = 1{\cdot}7\frac{T_{de}}{\tau}$$

$$T_I = \tau$$

$$T_D = T_{de}/2$$

As many processes are non-linear, optimum conditions may vary with load. In such cases the adjustment suitable for the least favourable conditions must be adopted.

9. Open-loop control

9.1. General

As explained in Chapter 1, sensing elements in open-loop controllers are not subjected to the consequence of their action. This has the disadvantage of not being able to shorten the time taken to eliminate a deviation purposely applying more corrective action than will eventually prove necessary, relying on an oportunity to apply counter-action in time to avoid excessive over- or undershoot. If, to this extent, the adjustment of open-loop controllers and elements may be more critical than in closed-loop control, this disadvantage is offset by the fact that, for the same reasons, instability is impossible.

Open-loop controllers or elements may have many different functions:

(*a*) They may be used for the control of processes not controllable by closed-loop methods;

(*b*) They may cause the set value of a closed-loop system to be reset according to a fixed schedule or in sympathy with an arbitrary variable.

(*c*) They may initiate action for protective purposes;

(*d*) They may improve the function of a secondary closed-loop controller by reducing its span or by resetting its control point (see Chapter 5).

9.1.1. Controlling the 'uncontrollable'

Summarising in a few sentences the main conclusions of earlier chapters, one might say: Apart from a large control range which is detrimental to the control quality under all circumstances, in processes with large τ and short T_{de} practically any controller is applicable. As T_{de} increases, controllers of the I type rapidly become useless. Two-step and P controllers, on the other hand, though their field of application is much wider, also become less satisfactory so that only P + I, P + D, and P + I + D, control remain. I controllers give excellent results when both τ and T_{de} are short.

There remain the processes with short τ and long T_{de} that might might conceivably be tackled by a P + I + D controller, but for which all the more simple controllers are useless. In such cases open-loop control may provide a cheap alternative. Take the case of a concrete floor in which heating coils have been laid as part of a space-heating installation. If it is controlled by a room sensing element, the temperature of the large mass of the floor will have to be raised before the ambient temperature can start to rise. This in itself is also a slow process, due to the small difference between floor and air temperatures $\theta_f - \theta_a$. In addition, when θ_a has finally reached the desired value, so much heat is stored in the floor that θ_a will continue to rise uncontrollably for a considerable time after the valve controlling the heat supply has been closed. Conditions are not quite as bad when θ_a passes the desired value on cooling down, but a considerable undershoot must be reckoned with before the downward trend can be arrested. This behaviour is conditioned by the nature of the plant and cannot be materially affected by any simple controller.

In such cases the application of a 'weather controller' to adjust flow temperature to a value commensurate with weather conditions may not be ideal, but has proved to be a great improvement on any direct control.

Overhead heating is often controlled by open-loop methods, sometimes for the same reasons, but sometimes on different grounds. The heat dissipation of heated ceilings is largely by radiation, and detectors sensitive to radiant heat in a similar way to the human body are not commercially available. As they react mainly or exclusively to the air temperature, they cannot prevent the effective temperature from rising to a level often too high for comfort.

A wholly different reason for applying open-loop methods is the following: In large buildings, no detector location can be found at which the heat requirements are characteristic for those of the building or zone as a whole. Alternatively, it may be desirable to impose indoor climatic conditions on all occupants so as to avoid discussion. In both cases open-loop control is indicated.

9.1.2. The resetting of γ_i

Several examples of the desired value of a process being affected by open-loop methods were given in Chapter 5: the raising of the set value in winter as outside temperature drops and, in summer, as it rises are cases in point. Considerations of comfort may also lead to raising γ_o as a function of ambient temperature in dew-point control plants. Here one temperature is influenced by another for the purpose of affecting a wholly different physical quantity – the relative humidity in the space controlled.

Carrying this alienation process a step further, in any time schedule control such as the day-night boost programme of many heating installations, of which industry provides innumerable similar examples, one physical quantity – temperature – is affected by a wholly different one – time. Inasmuch as the two are in no other way connected and could not normally influence one another in any way, this is a perfect example of open-loop control.

9.1.3. The protective function

An example of an open-loop element applied for protective purposes is a frost thermostat that, in the event of danger, initiates an optical signal, cuts out the fan and opens the heating valve; inasmuch as its contact mechanism cannot return to its 'safe' position once the danger has passed, there can be no feedback and the loop is broken.

A flame-stat, thermo-couple or photocell, installed with the object of protecting a plant against flame failure, is a further example, as well as high limit stats in floor heating, etc.

9.1.4. The reduction of the control range

It is frequently possible to measure one or more disturbances and to counteract their influence before they have been able to affect the controlled condition. As a consequence, the flow of energy to the process can be made to vary within narrower limits with impunity and the remaining disturbances can be dealt with even though the control range of any secondary controller is reduced. By thus reducing the gain, the control quality a is improved and simple controllers can be applied under conditions, in which, without precontrol, they would be unacceptable. Subdividing a single large boiler into a battery of smaller ones of which the individual units are taken into operation as a function of outside temperature keeps the span to a minimum under all load conditions, which is of special importance at low load. Even better results are obtained when fluctuations in internal demand are counteracted by a secondary controller which can only affect the heat production of the last boiler to be switched on (Fig. 9.1). A further example is given in Section 9.3.

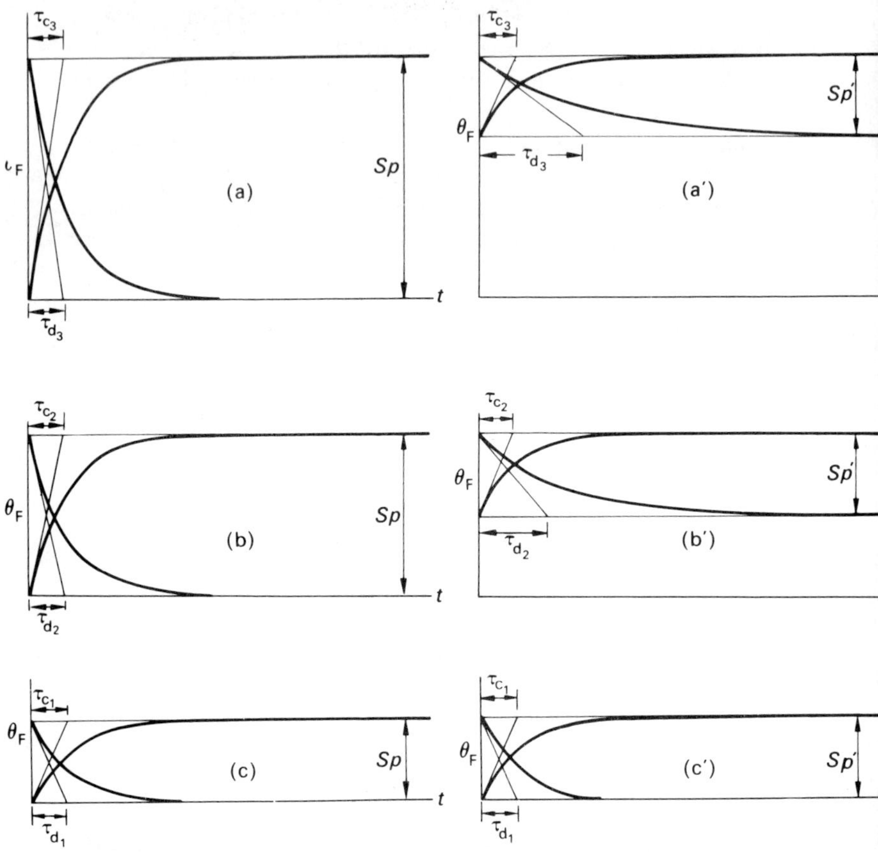

Fig. 9.1.
Reduction of control range in a heating installation in which the three boilers are switched consecutively as the heating load increases. Note charge and discharge time constants τ_c and τ_d when three, two and one of the burners are switched on and off together. (*a*), (*b*) and (*c*) and when only the last burner to be taken into operation is used for the adaption of supply to demand; the other(s) operating continuously. (*a'*, *b'* and *c'*) and the suffixes 1, 2, 3 indicate the number of burners in use.

9.2. Weather-sensitive controllers

Owing to their importance to the heating and air-conditioning field, a short section, possibly of less interest to the general reader, must be

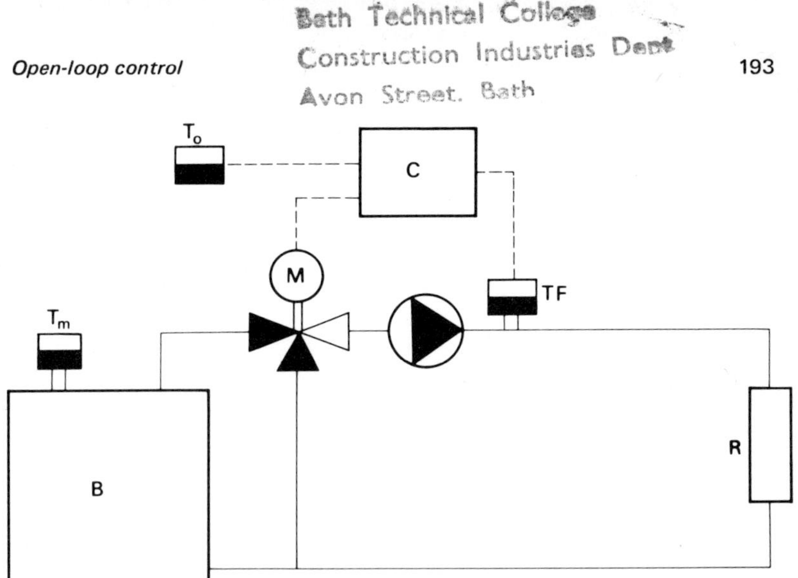

Fig. 9.2
Usual application of a compensator. B, Boiler, C, Controller, M, Motorised three-way mixing valve, R, Radiation, T_F, T_M, T_O, flow, maximum, and outside sensing elements respectively.

devoted to 'compensators', 'weather controllers' or whatever other fancy name these useful instruments have been given. Apart from a sprinkling of two-step controllers, they are overwhelmingly of the integral family, the set point being shifted upwards or downwards as climatic conditions cause the heating load to rise or drop.

The original practice of switching a burner on or off for the purpose of maintaining the desired flow temperature has become less and less acceptable if only on account of the danger of boiler corrosion at low loads, i.e. at low flow temperatures. Therefore, the modern solution is to run the boiler at a constant, high temperature to place a mixing valve in the supply to the installation and to control the flow temperature by changing the valve position (Fig. 9.2). Such devices are in general use both by themselves or as primary controllers, reducing the span for subsequent secondary controls.

9.2.1. Factors determining loss of heat

(*a*) All compensators are made to react to outside temperature, undoubtedly the main single influence on the heat requirements of an average building.

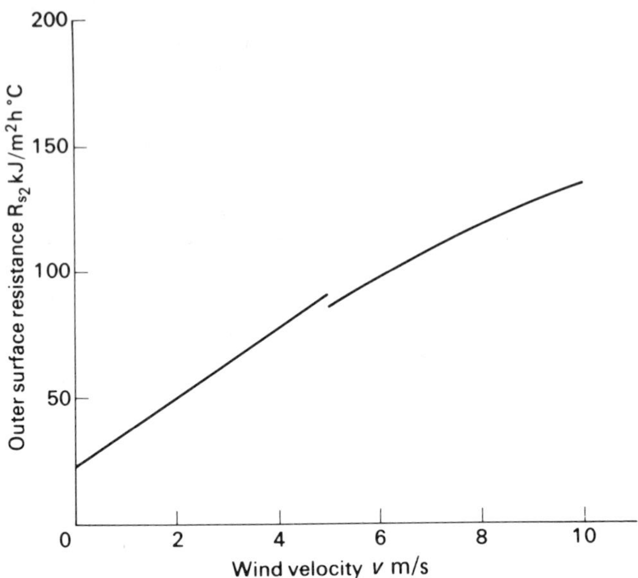

Fig. 9.3
The influence of wind velocity v parallel to the wall on the outer surface resistance, R_{s_2}, according to Jürges.[8]

(*b*) The second factor, wind (both temperature and velocity) is becoming more important as modern building methods favour skin type walls and large glass surfaces. The external surface conductance rises with wind velocity, v, according to Jürges[8] as quoted by Recknagel

$$\text{for } v < 5 \text{ m/s} \ \frac{1}{R_{s_2}} = (21 + 14{\cdot}2v) \text{ kJ/m}^2 \ 60\text{s}^\circ\text{C}$$

$$\text{for } v > 5 \text{ m/s} \ \frac{1}{R_{s_2}} = 25{\cdot}7\, v^{0{\cdot}72} \text{ kJ/m}^2 \ 60\text{s}^\circ\text{C}$$

If the wind flow is parallel to the wall (Fig. 9.3). As R_{s_2} is only one of the factors determining the U value, the other main features being the internal surface resistance $R_{s'}$ and the resistance of the structure

$$R = Lr$$

which is the thickness L times the resistivity, r, of the wall itself, the

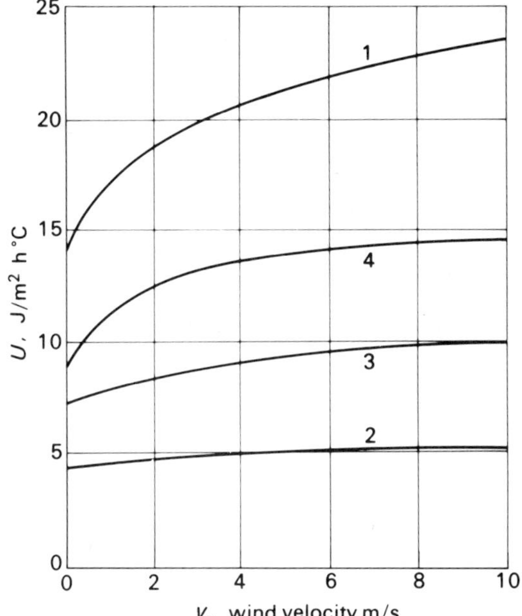

Fig. 9.4
The influence of the wind velocity, v, on the transmittance coefficient, U, is the greatest at low values of v. (1) for 4 mm glass; (2) for a wall of reasonable quality; (3) as for (2) with 25 per cent glass; (4) as for (2) with 50 per cent glass.

effect of the wind is reduced to the extent that $R_{s_1}F$ and Lr are larger. Yet the U values may increase by 10 per cent for a thick one and a half stone wall, and by 50 to 60 per cent for a 4 mm glass pane when the wind velocity changes from zero to 10 m/s; and by 8 and more than 40 per cent, respectively, when zero wind velocity rises to 4 m/s (Fig. 9.4).

The second effect of v is the increase in infiltration losses (Fig. 9.5), These reach their maximum when the wind direction is perpendicular to the wall, the increase being proportional to v, when the flow through the cracks is laminar, and to v^2 when it is turbulent, on the average therefore it is proportional to $v^{2/3}$. As the pressure increases with v^2 the heat loss is proportional to $v^{4/3}$. The cracks occur mainly around doors and windows when the latter can be opened and between wall and window frame when they cannot.

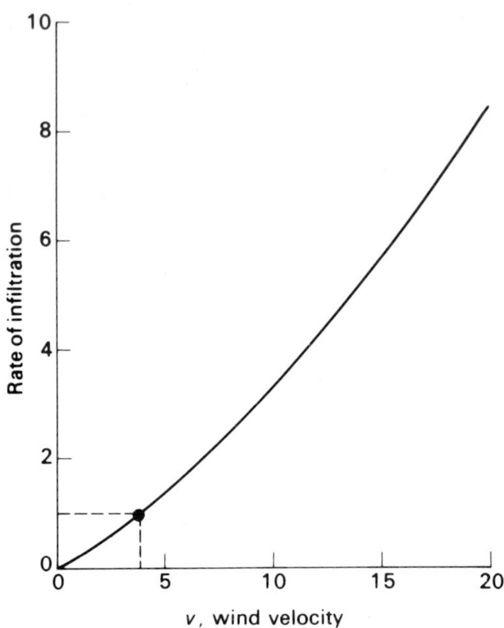

Fig. 9.5
The increase in rate of infiltration through cracks etc. as a function of wind velocity compared with infiltration at $v = 4$m/s according to Kopp.[9] The natural infiltration due to chimney effects etc. is neglected so that, in fact, losses at $v = 0$ m/s do not drop to zero.

The wind causes a pressure difference to arise over the building as a whole. As it is the air penetrating the wall at the luffside that displaces the outflow on the lee side, the resulting losses are profoundly affected by the resistance to be overcome by the air flowing through the structure from one side to the other. If a wall completely impenetrable to air flow could be interposed between the two sides, the losses due to infiltration could be reduced to a very small value. Therefore it is the air tightness of the building as a whole rather than that of the outside walls alone which determines the amount of displaced air and, therefore, the amount of cold air to be heated up. Similar to the wind-dependent transmission loss, the heat loss due to infiltration is proportional to the difference between wind and space temperatures. Together, the increase in heat loss due to wind can in extreme cases amount to a value comparable to transmission losses in calm weather.

(*c*) Considered as a loss, the heat gained by a building due to solar radiation is, of course a negative quantity. Its magnitude depends on the intensity of the radiation as well as on the angle of incidence both in a horizontal and a vertical plane. As short-wave radiation it largely penetrates glass into a building where, striking the floor and miscellaneous objects it is instantly converted to low temperature, long-wave radiation which is reflected by glass surfaces and, therefore cannot escape. Therefore, the resulting rise in space temperature also depends on the size of the window area.

Solar radiation striking and absorbed by the walls also contributes to the heat gain but to a lesser degree. It is largely converted to heat on the outer surface. Before it can penetrate into the mass of the wall and raise its temperature, part of it is carried off by the surrounding air, the more, the stronger the wind and the colder the weather. This is one reason why this type of heat gain is particularly noticeable in summer. Another notable feature is the delay with which it affects the temperature inside the building, making itself felt long after the time it was produced. This effect can only be countered by controls inside the enclosed space and will, for present purposes be neglected. There remains the window area, shading and orientation as the main factors determining the effect of solar radiation on space temperature in so far as it can be counteracted on the basis of a direct measurement of its intensity.

Even in our latitudes in the winter period, the heat gain in modern and especially high buildings can be such as to render them practically uninhabitable until a cooling plant capable of dealing with the effect is installed. All present-day compensators do is to turn off the heat. Although their use is therefore restricted to the period when there is heat to cut off, this, when all is said and done, is well worth doing.

(*d*) The last factor influencing heat loss is the moisture in and on the walls and windows which, coupled with tangential wind velocity, determines both their external surface conductance coefficient and – to the extent that the wall is porous – its conductance proper. Observing a thermometer in a hot-house at the onset of a thunderstorm, one can actually see the temperature drop as the rain beats on the glass enclosure. The same applied, albeit to a far smaller degree, to houses and office buildings. The resulting heat loss is largely a question of conduction although, when coupled with wind and non-saturated outside air, the loss of latent heat may also be clearly perceptible. Yet, compared to the other factors, the effect is but a minor one. Apart from a device called a rain detector, the resistance of which drops sharply when wetted, and which is used to close the windows of hot houses at the onset of a shower, this source of heat loss has, as yet, not been considered worth while to counteract.

The same applies to the increased conductance of porous walls when wet or even damp. In this respect an interesting phenomenon is reported from countries where frost usually occurs after a prolonged wet period. When moisture in the walls freezes, both evaporation and infiltration are suddenly reduced so that, instead of the heat loss increasing due to the drop in outside temperature, it may decrease due to better insulation.

Summarising, enough has been said to show the value of measuring solar radiation and wind velocity in addition to outside temperature as a measure of the heating load to be compensated.

9.2.2. Locating the outside sensing element

However much the compensator may have been perfected, it can do no more than convert the signals it receives from its sensing element into control action as described. The sensing element, in its turn can do no more than react to the influences to which it is subjected. Thus, its location should be such as to ensure that these influences correspond as accurately as possible to those determining the heat loss of the building as a whole or in part.

In this respect, any solution is bound to fall short of the ideal. Concerning the building as a whole, the orientation of walls and windows will be of major importance for the way in which the heat loss is affected by wind and sun, and a sensing element mounted on one of the facades is necessarily located incorrectly for the rooms adjoining the others. Under these circumstances the sensing element should be placed on the wall through which the heat loss is likely to be the largest, secondary controls being used to prevent rooms requiring less heat from becoming too warm. Larger buildings are usually sub-divided into zones corresponding to their orientation, and – in skyscrapers – to their height above the ground, each zone being fitted with its own compensator. In this case also, secondary controls cannot be dispensed with if internal heat gain etc. is at all important. Even if moderate standards of occupancy and lighting levels are assumed, the two combined may easily amount to one half of the total heat load and are certainly worth considering.

Sometimes it is recommended that the sensing element should measure the temperature of the outer wall surface rather than that of the air. In view of the method by which the heat losses are calculated this would not seem logical, more especially as the wall temperature is not only determined by weather conditions but also by transitory effects emanating from the inside of the building, such as changes in occupancy, lighting and degree of insulation (e.g. curtains). It is therefore preferable to measure the air temperature rather than that of the wall.

Also, though for reasons of accessibility it is advisable to locate the outside sensing element near to a window, it should never be affected by warm inside air escaping from the adjoining room when the window is opened. For similar reasons the neighbourhood of exhaust openings of ventilating systems is to be avoided.

9.3. Aspects of secondary control

The main reasons for adding a closed-loop controller to an open-loop control system have been discussed in earlier sections of this chapter. Inasmuch as the open-loop system was not applied specifically to improve operating conditions for secondary controllers, the latter are used to deal with disturbances not or not completely compensated by the outside sensing element, for example, on account of the practically inevitable inaccuracies in its adjustment. One important reason, not mentioned so far, is that the addition of a closed- to an open-loop element allows full advantage to be taken of internal heat sources for the maintenance of space temperature, even though they may be variable or intermittent. The simplest method of secondary control is to adjust the compensator to a value slightly in excess of requirements, and to switch over to the 'night' set point when the space temperature tends to exceed the desired value.

The advantages over a secondary on-off controller by which the roomstat operates a valve is that the span of the secondary controller is virtually constant especially at high load (Fig. 9.6). Let line (*a*) represent the flow temperature that would normally suffice to cover the heat requirements. These may increase to values corresponding to (*b*) under the influence of wind and can be reduced to (*c*) due to internal heat gain. The span of the secondary controller, Sp_c', is reduced to the difference in flow temperatures indicated by the lines (*d*) and (*e*), a small margin being added on either side for emergencies. As line (*e*) falls below the desired space temperature the latter takes over as the

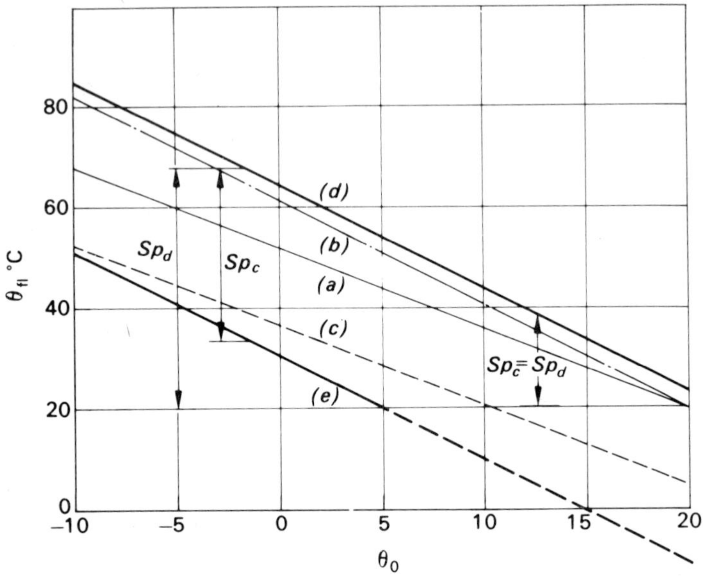

Fig. 9.6
Possible compensator flow-temperature curves corresponding to (*a*) normal heat load, (*b*) maximum heat load, (*c*) minimum heat load. The day characteristic could be adjusted to (*d*), the 'night' characteristic to (*e*), when the compensator is specifically used to reduce the span for secondary controllers.

lower limit. Thus, as the reduction of load would tend to aggravate the control problem, conditions are progressively improved by the decrease in span. Figure 9.7. shows schematically what happens at high (*a*) and at low load (*b*). Some of the control valves must be of the diverting type so that a minimum water velocity past the flow detector of Fig. 9.2 is maintained when all secondary valves happen to close simultaneously. The same applies when thermostatic radiator valves are used as secondary controllers. The advantage obtained by this method is not as great as might be expected.[13]

A very useful application of compensators is that in ventilation systems, where the air supply rated on the basis of the required number of air changes is insufficient to convey the heat necessary to maintain the space temperature. This function is carried out by radiators controlled by a compensator adjusted so as to maintain a somewhat

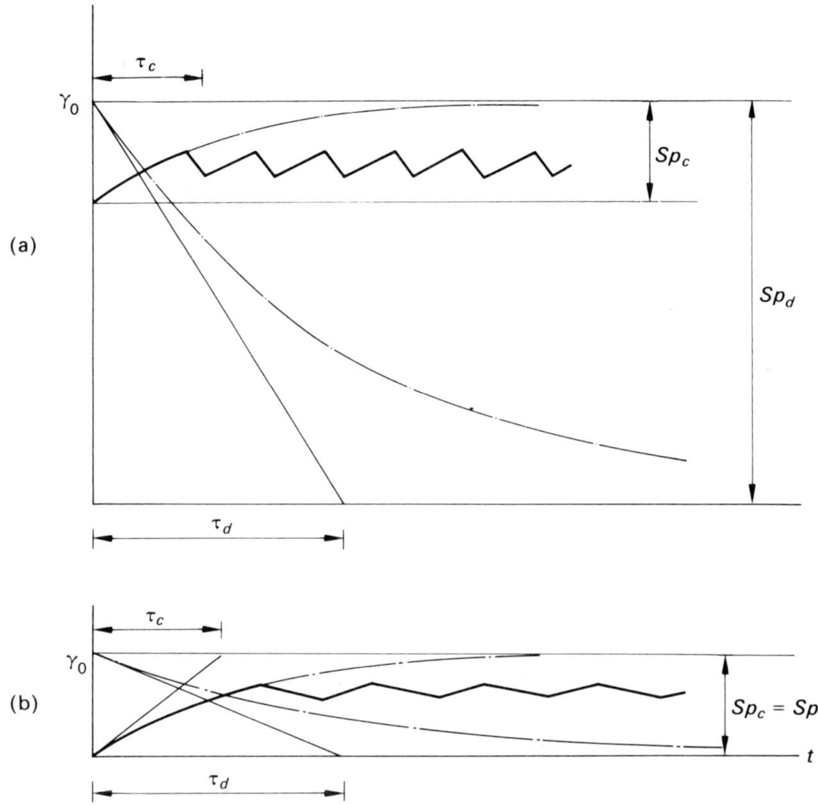

Fig. 9.7
The improvement in two-step control action due to reduction of control range by pre-control at (*a*) high load (*b*) low load.

lower temperature than ultimately required. The ventilation air is made to carry sufficient heat to top up the supply as necessary; the valve of the air heater battery being controlled by a simple modulating system, the sensing element of which is located in the space or exhaust duct.

9.4. Some compensator types

Many types of compensator have been developed in the past and have operated satisfactorily within the limits set by their design. The present

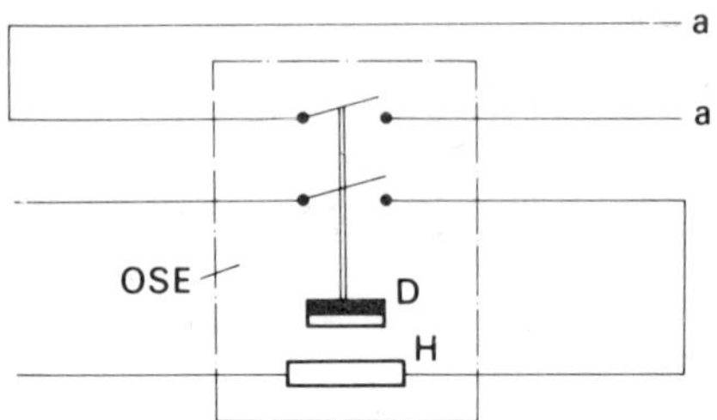

Fig. 9.8
A superseded type of compensator in which a thermostat D maintains the same temperature in the outside sensing element as in the house, the heat supply to which it controls by a second contact. H, electric heater.

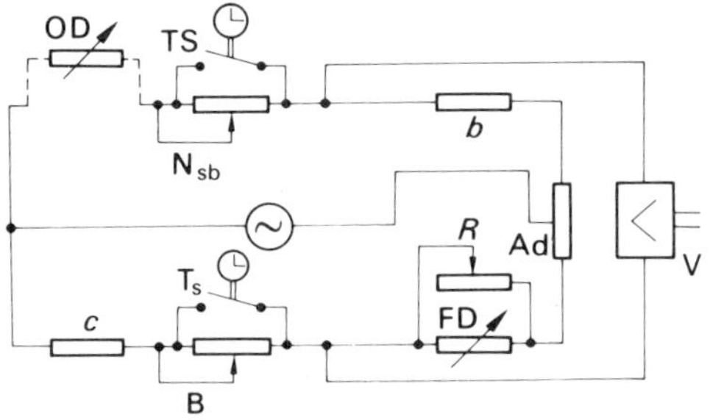

Fig. 9.9
Simple electronic compensator bridge circuit as described in the text.

tendency, however is for the self-operated purely electric and hydraulic types to be superseded by electronic and, in larger installations, pneumatic versions. Their mode of control is overwhelmingly of the floating or integral mode as, owing to the extremely short exponential lag of the mixing process, relatively large proportional bands would be required if, with a proportional controller, stability is to be achieved in the presence of effective dead time. Yet the proportional mode is usually adhered to in pneumatic versions for financial reasons.

The simplest of all is no doubt the electro-mechanical type used as an example in Chapter 1 (Fig. 1.4). Here the rate of change in flow temperature as a function of outside temperature is fixed, being determined by the volume ratio of the two detecting phials. Pneumatic compensators are invariably of the master-submaster type, a subject to be dealt with in detail in Chapter 10.

Another simple compensating system, shown in Fig. 9.8, employs a simulator. The outside sensing element *OSE* is fitted with an electric heater controlled by an on-off thermostat and rated in such a way that the ratio of the heat it supplies to compensate for the heat loss of the housing at various loads is as near as possible the same as that of the building it serves when the same temperature is maintained in both. A second contact linked to the first, is then used to control the heat supply to the house, which will, if all is well, maintain its temperature at the required level. Though the compensator described is obsolete, the principle of using a simulator in control circuits is encountered.

The bridge circuits of electronic versions differ only in detail from those described in Chapter 5. In the very much simplified method shown in Fig. 9.9, the outside and flow sensing elements *OD* and *FD* are placed in opposite bridge arms; as the resistance of *OD* drops, i.e. as it gets colder, that of *FD* must increase, i.e. the flow temperature must rise to maintain balance. The adjustment of the various temperatures is carried out by means of a single potentiometer Ad, whereas the ratio, i.e. the flow temperature rise for a 1° C fall in outside temperature, is effected by resistance *R* bridging a flow detector *FD* the use of which will necessitate a readjustment of Ad. Boost B and night set-back Nsb are obtained by adjustable series resistances shorted out by time switch contacts when necessary.

In the more comprehensive type, shown in Fig. 9.10, each element influencing the heat loss, such as outside temperature θ_0, wind velocity and temperature, W and solar radiation, S. is given its own bridge. The sum of the signals generated by outside influences is compared with the sum of those appertaining to the flow temperature, the difference between the two being offered to the amplifier which will initiate appropriate control action. The most simple of the bridges and the ones to which all others relate, is number 3 producing an output signal mirroring outside temperature θ_0. The output of solar bridge 4 is negative with respect to that of θ_0 and wind. The element exposed to the sun is therefore located at (*c*) whereas at (*b*) a temperature sensitive element compensates for changes in ambient temperature. In the wind bridge 2, the temperature-sensitive element is heated by the output of bridge 1 in such a way that, in calm weather it is held at a reference temperature of 20°C. As the wind rises it is cooled in proportion to both wind velocity and temperature. The sum of the output of bridges 2, 3 and 4 is applied to a winding of the input transformer. The main opposing signal comes from bridge 6 in which the flow temperature is

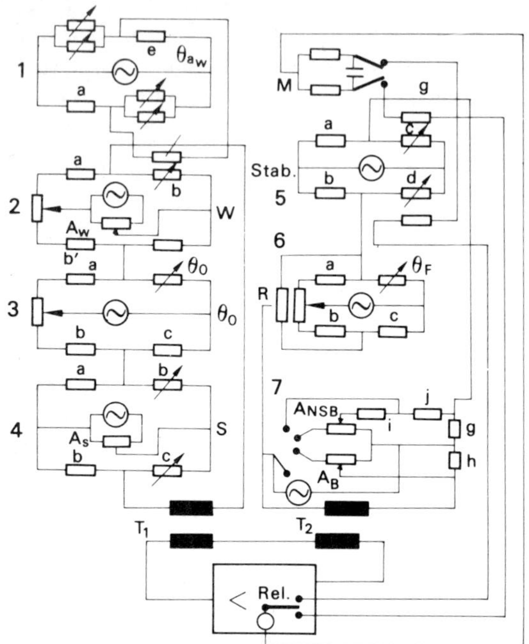

Fig. 9.10
More sophisticated bridge circuits compensating for solar radiation, wind velocity and temperature in addition to outside temperature (*Satchwell*).

measured. The ratio between flow and outside temperature changes being adjusted by potentiometer R over the bridge. Stability is ensured by thermal feedback bridge 5 similar to that described in the text to Fig. 3.10.

The arrangement at 7 serves to provide night set back Nsb and boost, the latter being remarkable for changing with load so as to keep the time required to restore day temperature after the night at a constant value.

9.5. The adjustment of compensators

It is unfortunate that compensators usually have to be adjusted under unsuitable circumstances. They depend to a greater extent than closed-loop controllers on the characteristics of the building they are to control and of its heating installation. In new buildings these are factors about which no one can give information based on actual experience, and those responsible for the design are seldom present to give what information is available. New buildings need more heat in the early months than later on as they have to be dried out. Finally, weather conditions in summer and early autumn are not the most suitable for trial and error methods. Therefore, though an experienced heating or control engineer does not need long to adjust an installation to give reasonable results, one or more further visits under different weather conditions are usually necessary if any degree of accuracy is desired. Frequently the necessary instructions can later be given on obtaining answers to such questions as: 'Is the temperature correct both in mild and in cold weather?, when it is windy or calm?, sunny or clouded?' That is, if an observant and willing occupier of the building can be found without whose assistance no satisfactory results can be expected anyway.

Yet, certain general features are of importance in all cases, such as the outside and flow temperatures under design conditions. This gives the highest point in the characteristics and, together with the lowest point, 20°C flow at 20°C outside temperature for most radiator installations with forced circulation, gives the ratio or slope of the characteristic. In the case of air heaters, the minimum flow temperature is likely to be approximately 45°C, in installations with natural circulation the lowest temperature at which circulation takes place is frequently 30 to 40°C unless special precautions are taken.

If the compensator contains means to take the influence of sun and wind into account, it is usually the ratios of window to wall surfaces and of surface to volume that have to be taken into account as well as the *U* value of the walls. The degree of exposure and the orientation of the wall are of less importance than one might think; sun and wind elements in themselves compensate for wind and solar influences as far as they are present; a large influence allotted to them will provide no compensation unless radiation and wind actually occur. Thus, surprisingly, a compensator with its outside element placed on a northern wall with many large windows, and which controls a zone with the same aspect, must have its sun element adjusted as if the orientation were southerly.

The night set-back, referring to the flow temperature, should not be too small as the new level will be reached quickly and heat dissipation continue even though the space temperature desired during the night has not yet been reached. Unless the boost in itself is varied with weather conditions, it is usually made large so as to shorten the boost period as much as possible.

10. Two or more controllers influencing the same variable

10.1. Mutual interference

It is a golden rule *never to allow one and the same variable to be directly and continuously controlled by more than one independent controller*, as, otherwise, instability will invariably occur owing to mutual interference. Of course, this does not apply to controllers and the like that are inherently unstable, like those of the two- or multi-positional type. Thus the arrangement shown in Fig. 10.1, in which a number of local air heaters in the same large space each controlled by its own on-off thermostat can work quite satisfactorily, especially as the span of each heater is limited. The unavoidable mutual interference will cause the switching of the individual heaters to become more irregular than if one heater only were present, but on the whole the space temperature will be remarkably constant. The only point to watch is that the desired values should be adjusted to be exactly equal.

The quality would certainly have been less had the controllers been of a continuous type; and the amplitude of the hunting will generally be less the greater the number of units. Assuming that the variable as measured by one of the sensing elements is influenced by a random disturbance. The corrective action initiated may affect the second detector before the disturbance is noticeably affected. Thus its input signal will assume the opposite sign. The resulting action may increase the deviation of controller 1 and soon hunting is initiated that is difficult to stop, each controller opposing the corrective action of the other by opening its valve as the other is closed and vice versa. In certain, fortunately rare, cases the use of more than one controller for a single variable is unavoidable, in which case the following considerations may be helpful.

The closer the control constants of the two loops resemble one another the larger the amplitude and the longer the period of the oscillation are likely to be. One method of reducing the effects of mutual interference, therefore, is to make T_{de} and τ as different as

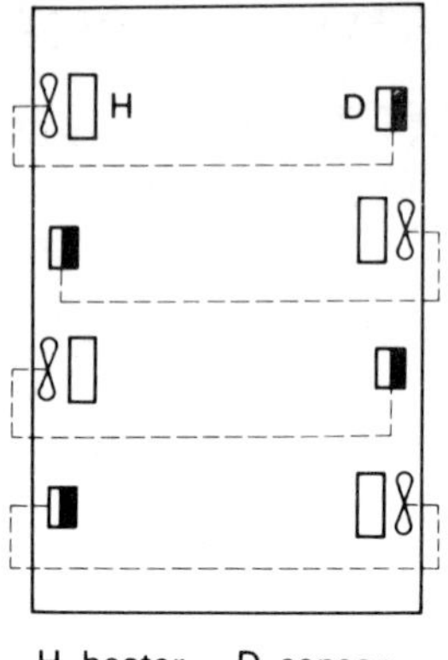

Fig. 10.1
A number of air heaters mounted in the same space each controlled by its own on-off thermostat.

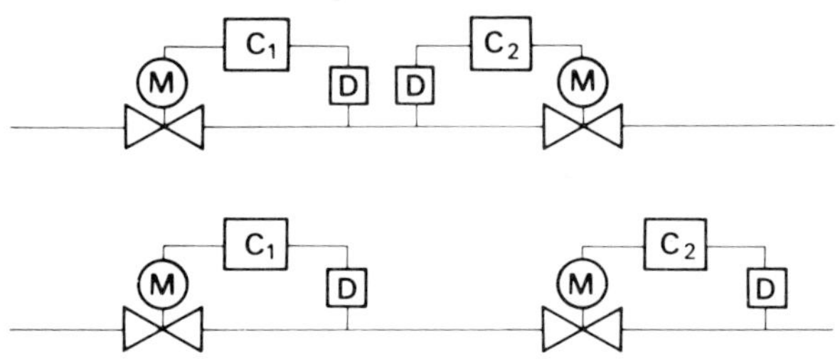

Fig. 10.2
Pressure control in a pipeline; (*a*) arrangement causing instability; (*b*) improved arrangement.

possible by locating the sensing elements accordingly. If the control range of each controller is not a given entity, as is usually the case, it will help to make these different also. This, together with the PB adjustment etc., can cause the amplification of one controller to be as much larger than that of the other as circumstances permit.

However, situations in which these measures become necessary should never have arisen. The variable controlled by each of the controllers should have been different. Thus the arrangement of Fig. 10.2(*a*) is quite wrong, that shown at (*b*), though not ideal, is much better. If a radiator installation is supplemented by an air heater for ventilation purposes, it is inadvisable to control both from the space. It is better to

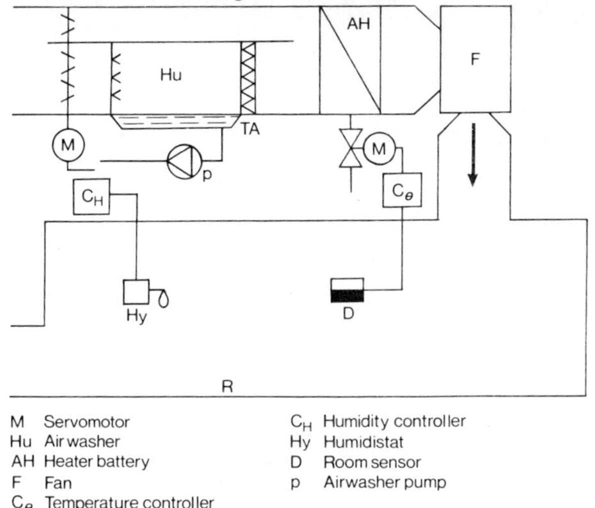

Fig. 10.3
Mutual interference of the temperature and relative humidity control schemes.

keep the discharge air temperature constant controlling the radiators from space, or to control the latter by means of a compensator and mount the sensing element for the heater in the space or in the exhaust duct. Separating the controlled conditions is not always effective. Figure 10.3 shows an air-conditioning installation in which both temperature and humidity are controlled from the space. When the air is too dry, more air will be made to pass through the humidifying chamber where its temperature drops as its latent heat rises with increasing moisture content. This, in turn, causes the discharge air temperature to drop which will, by affecting space temperature, satisfy the hygrostat even though the absolute moisture content required has not yet been reached. The lower room temperature, however, will cause the air heater valve to be opened up. If its proportional band is wide, so that the corrective action is aperiodic, all is well and the new stable situation will come about after a few more alternative steps by the two controllers. If, on the other hand, the amplification factor of both controllers is fairly high, the cycling of the one will cause the other to cycle in counter-phase, and the system as a whole will hunt uncontrollably.

One means of overcoming such difficulties is, therefore, to reduce

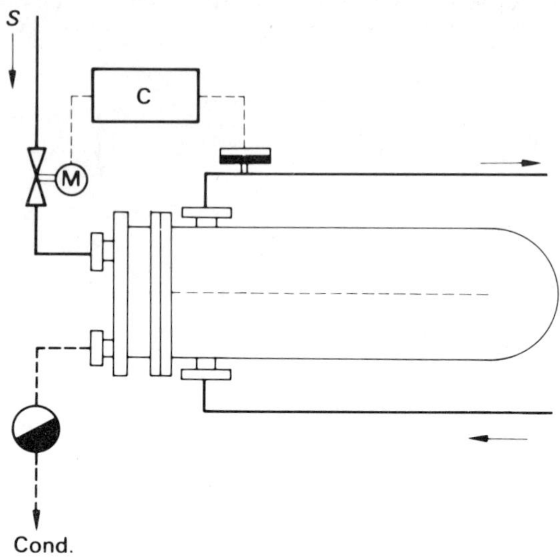

Fig. 10.4
Interference by the steam trap with the flow temperature control of a non-storage calorifier.

the amplification factor in at least one of the control circuits. In addition T_{de} and τ can be diversified, for instance by measuring the temperature in the exhaust duct and bringing the hygrostat nearer to the discharge grill.

Where possible, the control range of either or both should be reduced to a minimum. Sometimes it will be possible to restrict the stroke of the air damper so as to maintain a minimum degree of humidification, or some means of humidification can be placed in the space and operated continuously during periods of high load. In dew-point installations, a master-submaster arrangement (dealt with later in this chapter) will give excellent results.

There are other examples of mutual interference which, if not always serious, may prove to be puzzling or disquieting. For example, the large differential of a boiler stat may cause periodic movement of a mixing valve supplied with hot water.

A notable example is the apparent instability of the secondary flow temperature control in non-storage calorifiers heated by steam which is actually due to the discontinuous operation of the steam trap (Fig. 10.4). In its closed position the trap will cause the condensate to flood a steadily increasing part of the heating surface. The flowstat, detecting the resulting drop in heat transfer, will cause the steam valve to open further. The secondary steam pressure rises as do the condensation temperature and the transfer of heat. At a given moment the trap opens, the original heating surface is suddenly restored to usefulness, the flow temperature rises and the steam valve has to close hurriedly. Apart from the resulting ripple in the secondary flow temperature, which may, or may not be objectionable, the wear on valve and actuator makes the use of a ball float type of trap, which works continuously, advisable. Admittedly, this trap is a proportional controller, practically independent of the temperature control loop. It might therefore be expected to cause interference of a different kind. That this does not happen is due to the practical absence of dead time as a result of which the float operates aperiodically under nearly all conditions; the control constants of the two control systems differ too much for more serious trouble to occur.

10.2. Co-operation of two or more controllers: cascade and master-submaster control

A widely used arrangement in which two controllers are made to support rather than to oppose one another is called master-submaster control. Its official name is 'cascade' control, according to B.S. 1523, a system in which one controller provides the command signal to one or more other controllers. The variables to which the individual sensing elements react must be related but are never identical.

As the block diagram of Fig. 10.5 shows, cascade control involves at least two control loops. The secondary, short loop control circuit of the submaster involves a process in which a variable is kept constant by a separate controller, a final control element and a sensing element. All these elements together serve as a final control element to the master in the main long loop control circuit in which the primary variable is to be maintained at a constant value. This latter variable provides the signal that resets the desired value to which the submaster controls its variable.

An example will clarify what is meant. Let an air heater be controlled in the conventional way by a sensing element in the exhaust duct (Fig. 10.6(*a*)). If less heat is required the valve will close a little and the mean heater temperature will drop in accordance with its time constant (*b*). Normally, the valve will only partially close so that a certain amount

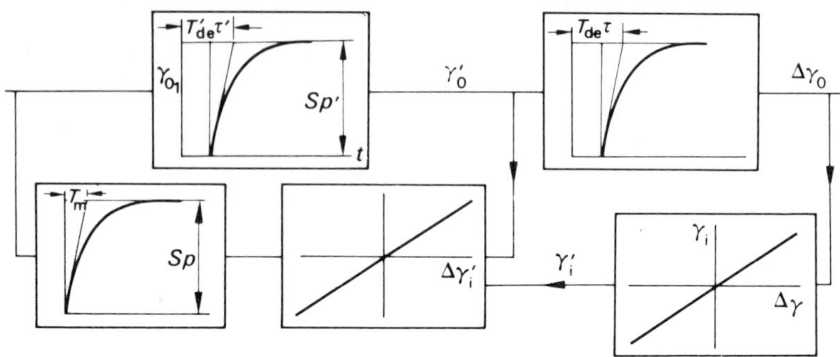

Fig. 10.5
Block diagram of a cascade control circuit. The three blocks on the left refer to the short secondary control loop.

of hot water will continue to flow through the heater and thus delay the moment at which the average temperature desired by the detector is reached. This slow reaction will distort the response curve as measured in the exhaust (*c*), increasing the lag T_{de} inherent in the process, thus reducing the accuracy and speed with which θ_e can be maintained at the required level (Fig. 10.6(*c*) and (*f*).

Now consider the action of a cascade control system (Fig. 10.6(*d*)) in which the sensing element reduces the desired value of the discharges at γ'_i, to an extent proportional to the deviation. As a result the valve will close frequently and be kept closed until the desired discharge temperature has been reached. The lag inherent in the main circuit T_{de} is scarcely affected (*f*) to the benefit of control quality. Reduction of time lags, therefore is one of the reasons for applying cascade control.

The same example can be used to demonstrate other desirable features. Certain disturbances and non-linearities etc. are removed near their source and well before they can influence the primary controlled variable. In Fig. 10.6 such disturbances may be changes in outside air temperature or quantity; variations in flow temperature at the control valve, the non-linearities caused by an oversized heater or/and control valve all of which could affect θ_d and therefore θ_e if allowed to pass the submaster unchecked. In fact, all disturbances occurring before the air reaches the thermostat are taken care of by the submaster which, moreover, operating in a very short loop can attain an excellent control quality. Relieved of a large part of its earlier duties, the control range of the master control circuit is effectively reduced, so that here

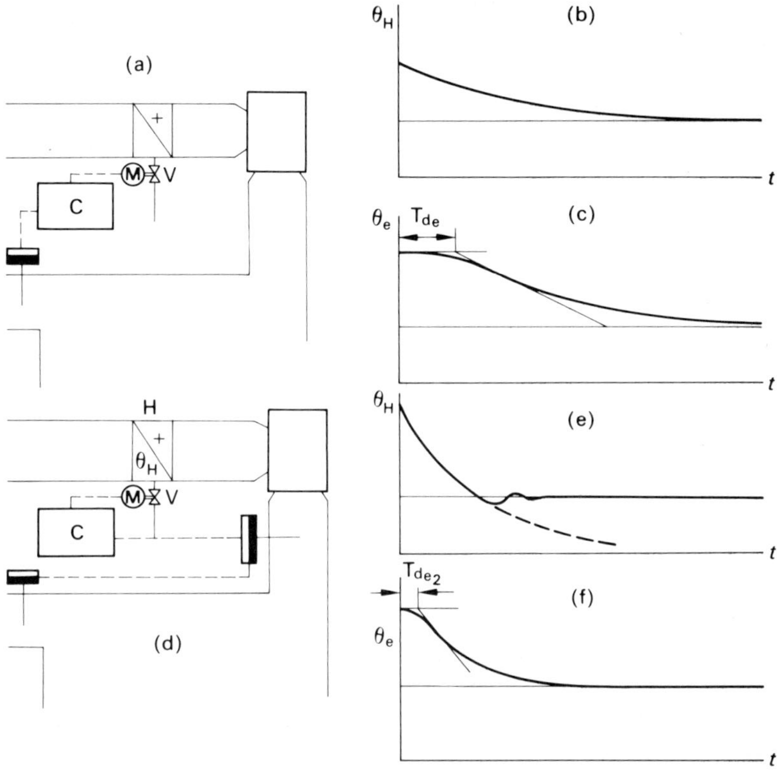

Fig. 10.6
Conventional (*a*) and cascade (*d*) control systems maintaining the exhaust air temperature θ_e at a constant level. In the second case the devised value θ_d of the dischårge sensing element is reset in accordance with the temperature θ_e. Thus the mean heater temperature θ_H changes more quickly than in the conventional way (*b*), (*e*) and the effective dead time T_{de} is reduced (*c*), (*f*).

also, notwithstanding the less favourable control constants, the control quality can be materially improved. This of course, does not mean that the capacity of the valve and air heater can be reduced; the control range of the system as a whole is not affected.

A third object of cascade control may be to cause one variable to change in direct proportion to another. Thus many of the open-loop schemes discussed in the previous chapter can be included in this group as cascade controllers. The example shown in Fig. 10.7. refers to the secondary control in a compensator circuit where the control point

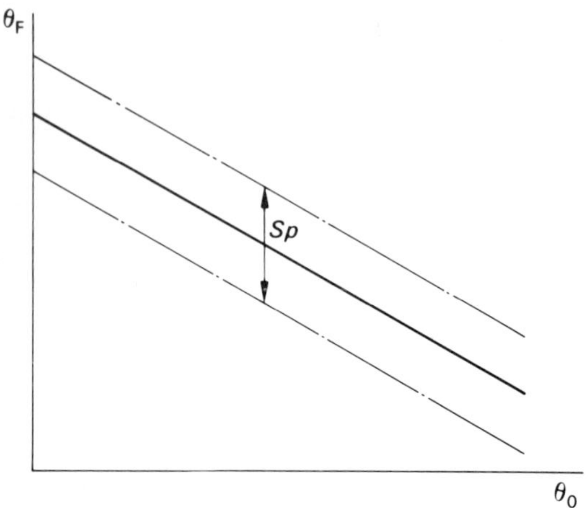

Fig. 10.7
Resetting the devised value θ_F of a weather controller from space (θ_3); Sp = control range of the space temperature control loop; θ_0 = outside temperature; θ_F = flow temperature.

of a compensator is shifted upwards and downwards within the band between the day and night desired values so as to counteract disturbances such as changes in occupancy, heat gain from artificial lighting, etc. Figure 10.8 shows a further example in which the temperature of a flow of water is raised by injection of steam. Fluctuations in water pressure etc. may influence the rate of flow, and steam pressure fluctuations the amount of steam injected; both factors will affect the ultimate flow temperature. The rate of reaction of this kind of process being extremely fast, even short lags have a disproportionately detrimental influence on control quality. By measuring the rate of flow in the water mains and using this signal to raise or drop the submaster control point and thus to open or close the steam valve, the flow of steam is adjusted to requirements before the new temperature front has reached the flow temperature detector which, as submaster, further modifies the valve position. Thus one of the two main disturbances is largely accounted for, and the span of the primary control loop is reduced in proportion. The same scheme can also serve as an example for the last main purpose of cascade control, namely, to keep a secondary variable within safe limits.

Should the water flow suddenly cease, the water in the secondary circuit may start to boil and thus damage subsequent equipment unless

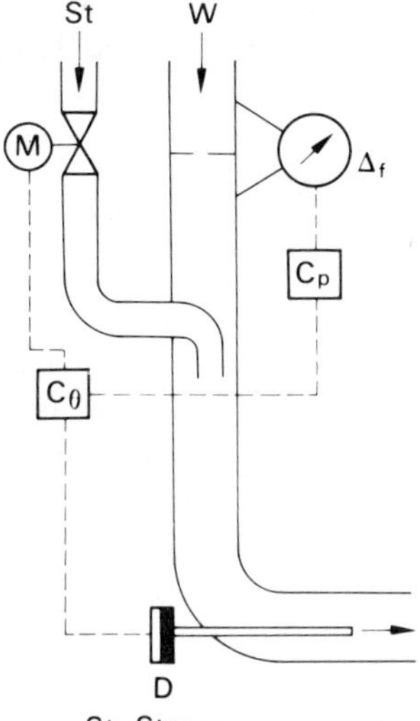

St Steam
W Water
M Servomotor
C_p Pressure controller
C_θ Temperature controller
D Sensor

Fig. 10.8
Scheme by which the rate of steam flow is made to conform to fluctuations in the amount of water passing, thus relieving the (main) temperature control circuit.

the valve is closed immediately. Should this equipment be sensitive to excessively low temperatures also, cascade control reduces the likelihood of a dangerous situation arising by its very nature.

10.2.1. The type of equipment

Though cascade control is not limited to any particular kind of controller, it falls so naturally to the pneumatic type that its application would seem to have given rise to the restriction in the B.S. definition which refers to controllers rather than to detectors.

Until the introduction of the transmitter system, separate pneumatic controllers, i.e. controllers without a detector as part of it, were rare, The output signal from the sensing element – an air pressure – could be used to reset the valve without interposition of further amplifiers, comparing elements etc. as these were incorporated in the sensing element.

To a certain extent this also applies to hydraulic controllers, but certainly not to electric and electronic types. Here, separate master and submaster controllers can frequently be replaced by a single master-submaster controller-amplifier, the input signal of which is subject to both master and submaster influences excerted in the same bridge circuit or in similar ones placed in series. Of the several arrangements of this kind discussed in earlier chapters the compensator was the most recent.

Master-submaster arrangements are widely used in pneumatic control systems. Here, the output signal of the master sensing element – read controllers – is applied directly to the submaster so as to change its set value. As both are normally of the proportional mode one may wonder how the proportional band of the submaster affects that of the system as a whole. Figure 10.9, referring to the arrangement of Fig. 10.6, illustrates what is meant. If the proportional band γ_{pm} of the master in the exhaust duct is adjusted to 2°C and its desired value to 20°C, it will reset the desired value of the submaster in this example to 40°C when θ_e = 19°C and to 20°C when θ_e = 21°C. The set value of the submaster, operative when θ_e = 20°C, will then equal 30°C. The submaster, also of the proportional mode, will have a proportional band of its own, say γ_{pam} = 6°C. Its desired value, therefore will depend upon load conditions, i.e. the sum total of the disturbances originating at some point before the air reaches the submaster. At low loads, the control point will approach the line, $i-d$ at high load it will approximate to the line f–g. In general it will be somwhere between these two lines. Should low load conditions prevail in both loops simultaneously, the discharge temperature will lie at point d, corresponding to 21°C in the example instead of to the 20°C envisaged; the discharge temperature will have to rise beyond 21°C before θ_d reaches the 20°C. Similarly, at high load θ_d = 37°C instead of the 40°C planned and θ_e will have to drop below 19°C if 40°C is to be reached. This means that γ_{pm} can increase or decrease as a result of load conditions in the submaster loop.

Owing to the congruence of triangles *bde* and *fbk* as well as *gch* and *cij*, $be = fb = gc = ci$. Furthermore, triangle *bde* is similar to *abc*, and therefore

$$be = \frac{bd}{ac}\, ab \quad \text{or}$$

$$\frac{\Delta\gamma_p}{2} = \frac{\frac{1}{2}\gamma_{psm}}{S_{psm}}\, \gamma_{pm}$$

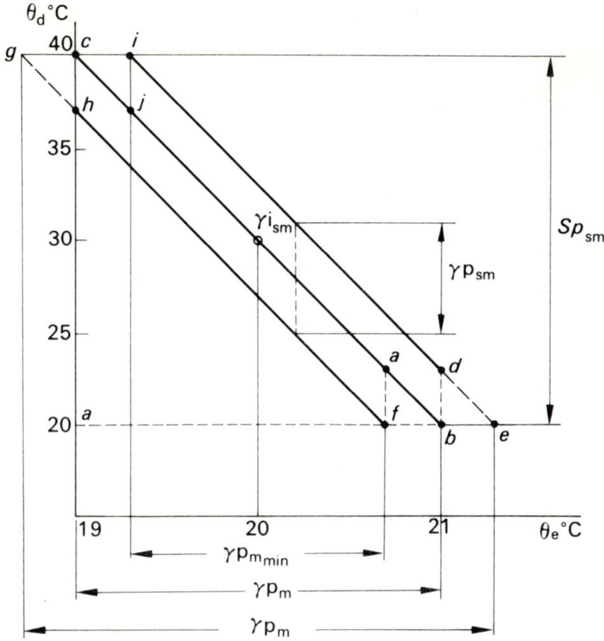

Fig. 10.9
The effective proportional band of a master-submaster control system in which both controllers are of the proportional type. θ_d, discharge temperature; θ_e, exhaust temperature; γ_{PM}, γ_{PMmin}, γ_{PMmax} the mean, minimum and maximum proportional bands of the master; γ_{Psm} that of the submaster.

In the example $\Delta PB = 2 \times \frac{6}{2} \times 2 \times \frac{1}{20} = 0{\cdot}6°C$. Therefore, the limiting values of the master-submaster arrangement as shown equal $2 \pm 0{\cdot}3$ or 1·7 and 2·3°C.

Theoretically it is possible for the γ_p to remain constant but for the desired value to shift between 19·7 and 20·3°C. It should be noted that the variability of γ_i and γ_p are due to the interaction of the two proportional systems. Pneumatic submasters of the integral family are relatively rare but quite common as electronic controllers (Fig. 10.10).

In this case γ_{pm} is not affected by conditions in the secondary loop, and γ_i also remains constant. However, the signal of electronic sensing elements is not restricted. Thus, if under severe load conditions the master temperature drops below the 19°C of Fig. 10.10, it will call for

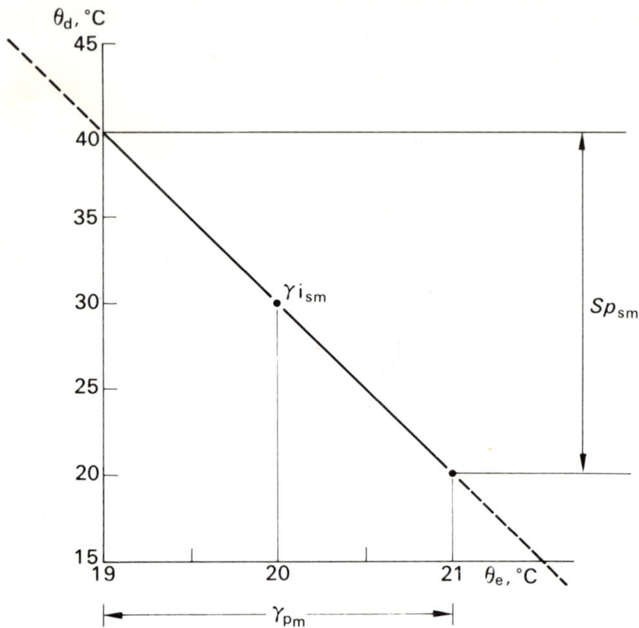

Fig. 10.10
If the submaster in an electronic master-submaster system is of the floating type, the submaster does not affect the PB of the master controller though the normal limits may be exceeded for other reasons. Symbols as in Fig. 10.9.

a higher discharge temperature than the 40°C of the example and, if the heater has any excess capacity, it will get it. Also if, owing to an internal heat gain larger than envisaged, the master detects a temperature higher than 21°C and if the air temperature before the heater happens to be lower than 20°C of the example, the discharge temperature will drop below that value. If either is undesirable, limiting devices will have to be incorporated to restrict θ_d to acceptable limits.

References

1. British Standard 1523, Glossary of Terms used in Automatic Controlling and Regulating Systems, Section 1. Process and Kinetic Control. British Standards Institution, London, 1954.
2. *ibid.*, Section 5, Components of servo-mechanisms.
3. *Automatic Control Terminology*, American Society of Mechanical Engineers, 1954.
4. K.L. CHIEN, J.A. HRONES and J.B. RESWICK, 'On the Automatic Control of Passive Systems,' *Trans. ASME,* **74,** 1952, pp. 175–85.
5. E. HECK, 'Regelkreise in der Klimatechnik und ihre Stabilisierung,' *Heizung-Lüftung-Haustechnik* **15,** 1964, April, pp. 125–9.
6. B. JUNKER,'Die regeltechnischen Grundlagen der Anwendung selbsttätiger Regler in der Heizung- und Klimatechnik' *Heizung-Lüftung-Haustechnik*, **7**, 1956, no. 11, pp. 177–86. An English version of this text is obtainable as *Sauter Bulletin*, no. 12-Re.
7. B. JUNKER, 'Die Regelgenauigkeit bei Klimaregelungen,' *Technische Rundschau*, **51**, 1959, no. 24. An English version of this text is obtainable as *Sauter Bulletin* no. 20-Re.
8. JURGES, as quoted in Recknagel-Sprenger, *Taschenbuch für Heizung, Lüftung und Klimatechnik*, p. 106, 53rd impression, R. Oldenburg, München, 1964.
9. L. KOPP, *Die Wasserheizung,* Springer Verlag, Berlin, 1958.
10. J. KROMWIJK, 'Het gedrag van graduele temperatuurregelaars in installaties voor verwarming en ventilatie,' *Verwarming en Ventilatie*, April 1954, pp. 157–70.
11. F. PIWINGER, *Regelungstechnik für Praktiker*, VDI-Verlag Düsseldorf, 1966.
12. M. SYRBE, 'Über die optimale Einstellung von Reglern auf Grund geometrischer Daten der Regelstrecken-Frequenzgänge' *Regelungstechnik*, **1,** 7, p. 160.
13. W.H. WOLSEY, *Control Valves*, Hutchinson Educational (in preparation).
14. J.G. ZIEGLER and N.B. NICHOLS, 'Optimum Settings for Automatic Controllers,' *Trans. ASME*, 1942, p. 759.
15. J.G. ZIEGLER, 'Cascade control systems,' *Canadian Chemical Processing*, October 1955.'